Aids to Forensic Medicine and Toxicology

W. G. Aitchison Robertson

Aids to Forensic Medicine and Toxicology

Copyright © 2022 Indo-European Publishing

All rights reserved

The present edition is a reproduction of previous publication of this classic work. Minor typographical errors may have been corrected without note; however, for an authentic reading experience the spelling, punctuation, and capitalization have been retained from the original text.

ISBN: 978-1-64439-605-6

CONTENTS

PART I

FORENSIC MEDICINE

I	Crimes	1
II	Medical Evidence	2
III	Personal Identity	10
IV	Examination of Persons found Dead	12
V	Modes of Sudden Death	12
VI	Signs of Death	15
VII	Death from Anæsthetics, etc.	19
VIII	Presumption of Death; Survivorship	20
IX	Assaults, Murder, Manslaughter, etc.	20
X	Wounds and Mechanical Injuries	21
XI	Contused Wounds, etc.	22
XII	Incised Wounds	22
XIII	Gunshot Wounds	24
XIV	Wounds of Various Parts of the Body	26
XV	Detection of Blood-Stains, etc.	30
XVI	Death by Suffocation	34
XVII	Death by Hanging	35
XVIII	Death by Strangulation	35
XIX	Death by Drowning	36
XX	Death from Starvation	38
XXI	Death from Lightning and Electricity	39
XXII	Death from Cold or Heat	40
XXIII	Pregnancy	40
XXIV	Delivery	41
XXV	Fœticide or Criminal Abortion	43
XXVI	Infanticide	44
XXVII	Evidences of Live-Birth	47
XXVIII	Cause of Death in the Fœtus	50
XXIX	Duration of Pregnancy	51
XXX	Viability of Children	52
XXXI	Legitimacy	53

XXXII	Superfœtation	54
XXXIII	Inheritance	55
XXXIV	Impotence and Sterility	55
XXXV	Rape	57
XXXVI	Unnatural Offences	60
XXXVII	Blackmailing	62
XXXVIII	Marriage and Divorce	62
XXXIX	Feigned Diseases	65
XL	Mental Unsoundness	69
XLI	Idiocy, Imbecility, Cretinism	70
XLII	Dementia	72
XLIII	Mania, Lucid Intervals, Undue Influence, Responsibility, etc.	73
XLIV	Examination of Persons of Unsound Mind	78
XLV	Inebriates Acts	80

PART II

TOXICOLOGY

I	Definition of a Poison	82
II	Scheduled Poisons	82
III	Classification of Poisons	85
IV	Evidence of Poisoning	87
V	Symptoms and Post-Mortem Appearances of Different Classes of Poisons	88
VI	Duty of Practitioner in Supposed Case of Poisoning	91
VII	Treatment of Poisoning	92
VIII	Detection of Poison	94
IX	The Mineral Acids	97
X	Sulphuric Acid	98
XI	Nitric Acid	99
XII	Hydrochloric Acid	100
XIII	Oxalic Acid	101
XIV	Carbolic Acid	103
XV	Potash, Soda, and Ammonia	104

XVI	Nitrate of Potassium, etc.	106
XVII	Potassium Salts, etc.	107
XVIII	Barium Salts	108
XIX	Iodine—Iodide of Potassium	109
XX	Phosphorus	110
XXI	Arsenic and its Preparations	111
XXII	Antimony and its Preparations	116
XXIII	Mercury and its Preparations	118
XXIV	Lead and its Preparations	120
XXV	Copper and its Preparations	122
XXVI	Zinc, Silver, Bismuth, and Chromium	123
XXVII	Gaseous Poisons	125
XXVIII	Vegetable Irritants	128
XXIX	Opium and Morphine	129
XXX	Belladonna, Hyoscyamus, and Stramonium	132
XXXI	Cocaine	133
XXXII	Camphor	134
XXXIII	Tetrachlorethane	135
XXXIV	Alcohol, Ether, and Chloroform	135
XXXV	Chloral Hydrate	140
XXXVI	Petroleum and Paraffin Oil	141
XXXVII	Antipyrine, Antefebrin, Phenacetin, and Aniline	142
XXXVIII	Sulphonal, Trional, Tetronal, Veronal, Paraldehyde	144
XXXIX	Conium and Calabar Bean	145
XL	Tobacco and Lobelia	146
XLI	Hydrocyanic Acid	147
XLII	Aconite	150
XLIII	Digitalis	151
XLIV	Nux Vomica, Strychnine, and Brucine	152
XLV	Cantharides	154
XLVI	Abortifacients	155
XLVII	Poisonous Fungi and Toxic Foods	156
XLVIII	Ptomaines or Cadaveric Alkaloids	158

PREFACE TO NINTH EDITION

I trust that, having thoroughly revised the "Aids to Forensic Medicine," it may prove as useful to students preparing for examination in the future as it has been in the past.

W.G. AITCHISON ROBERTSON.
Surgeons' Hall,
Edinburgh,
November, 1921

PREFACE TO EIGHTH EDITION

This work of the late Dr. William Murrell having met with such a large measure of success, the publishers thought it would be well to bring out a new edition, and invited me to revise the last impression.

This I have done, and while retaining Dr. Murrell's text closely, I have made large additions, in order to bring the "Aids" up to present requirements. I have also rearranged the matter with the object of making the various sections more consecutive than they were previously.

W.G. AITCHISON ROBERTSON
Surgeons' Hall,
Edinburgh.
June, 1914

PART I

FORENSIC MEDICINE

I

CRIMES

Forensic medicine is also called Medical Jurisprudence or Legal Medicine, and includes all questions which bring medical matters into relation with the law. It deals, therefore, with (1) crimes and (2) civil injuries.

1. A crime is the voluntary act of a person of sound mind harmful to others and also unjust. No act is a crime unless it is plainly forbidden by law. To constitute a crime, two circumstances are necessary to be proved—(a) that the act has been committed, (b) that a guilty mind or malice was present. The act may be one of omission or of commission. Every person who commits a crime may be punished, unless he is under the age of seven years, is insane, or has been made to commit it under compulsion.

Crimes are divided into misdemeanours and felonies. The distinction is not very definite, but, as a rule, the former are less serious forms of crime, and are punishable with a term of imprisonment, generally under two years; while felonies comprise the more serious charges, as murder, manslaughter, rape, which involve the capital sentence or long terms of imprisonment.

An offence is a trivial breach of the criminal law, and is punishable on summary conviction before a magistrate or justices only, while the more serious crimes (indictable offences) must be tried before a jury.

2. Civil injuries differ from crimes in that the former are compensated by damages awarded, while the latter are punished; any person, whether injured or not, may prosecute for a crime, while only the sufferer can sue for a civil injury. The Crown may remit punishment for a crime, but not for a civil injury.

II

MEDICAL EVIDENCE

On being called, the medical witness enters the witness-box and takes the oath. This is very generally done by uplifting the right hand and repeating the oath (Scottish form), or by kissing the Bible, or by making a solemn affirmation.

1. He may be called to give ordinary evidence as a common witness. Thus he may be asked to detail the facts of an accident which he has observed, and of the inferences he has deduced. This evidence is what any lay observer might be asked.

2. Expert Witness.—On the other hand, he may be examined on matters of a technical or professional character. The medical man then gives evidence of a skilled or expert nature. He may be asked his opinion on certain facts narrated—e.g., if a certain wound would be immediately fatal. Again, he may be asked whether he concurs with opinions held by other medical authorities.

In important cases specialists are often called to give evidence of a skilled nature. Thus the hospital surgeon, the nerve specialist, or the mental consultant may be served with a subpœna to appear at court on a certain date to give evidence. The evidence of such skilled observers will, it is supposed, carry greater weight with the jury than would the evidence of an ordinary practitioner.

Skilled witnesses may hear the evidence of ordinary witnesses in regard to the case in which they are to give evidence, and it is, indeed, better that they should understand the case thoroughly, but they are not usually allowed to hear the evidence of other expert witnesses.

In civil cases the medical witness should, previous to the trial, make an agreement with the solicitor who has called him with reference to the fee he is to receive. Before consenting to appear as a witness the practitioner should insist on having all the facts of the case put before him in writing. In this way only can he decide as to whether in his opinion the plaintiff or defendant is right as regards the medical evidence. If summoned by the side on which he thinks the medical testimony is correct, then it is his duty to consent to appear. If, however, he is of opinion that the medical evidence is clearly and correctly on the opposite side, then he ought to refuse to appear and give evidence; and, indeed, the lawyer would not desire his presence in the witness-box unless he could uphold the case.

Whether an expert witness who has no personal knowledge of the facts is bound to attend on a subpœna is a moot point. It would be safer for him to do so, and to explain to the judge before taking the oath that his memory has not been sufficiently 'refreshed.' The solicitor, if he desires his evidence, will probably see that the fee is forthcoming.

A witness may be subjected to three examinations: first, by the party on whose side he is engaged, which is called the 'examination in chief,' and in which he affords the basis for the next examination or 'cross-examination' by the opposite side. The third is the 're-examination' by his own side. In the first he merely gives a clear statement of facts or of his opinions. In the next his testimony is subjected to rigid examination in order to weaken his previous statements. In the third he is allowed to clear up any discrepancies in the cross-examination, but he must not introduce any new matter which would render him liable to another cross-examination.

The medical witness should answer questions put to him as clearly and as concisely as possible. He should make his statements in plain and simple language, avoiding as much as possible technical terms and figurative expressions, and should not quote authorities in support of his opinions.

An expert witness when giving evidence may refer to notes for the purpose of refreshing his memory, but only if the notes were taken by him at the time when the observations were made, or as soon after as practicable.

There are various courts in which a medical witness may be called on to give evidence:

1. **The Coroner's Court.**—When a coroner is informed that the dead body of a person is lying within his jurisdiction, and that there is reasonable cause to suspect that such person died either a violent or unnatural death, or died a sudden death of which the cause is unknown, he must summon a jury of not less than twelve men to investigate the matter—in other words, hold an inquest—and if the deceased had received medical treatment, the coroner may summon the medical attendant to give evidence. By the Coroners (Emergency Provisions) Act of 1917, the number of the jury has been cut down to a minimum of seven and a maximum of eleven men. By the Juries Act of 1918, the coroner has the power of holding a court without a jury if, in his discretion, it appears to be unnecessary. In charges of murder, manslaughter, deaths of prisoners in prison, inmates of asylums or inebriates' homes, or of infants in nursing homes, he must summon a jury. The coroner may be satisfied with the evidence as to the cause of a person's death, and may dispense with an inquest and grant a burial certificate.

Cases are notified to the coroner by the police, parish officer, any medical practitioner, registrar of deaths, or by any private individual.

Witnesses, having been cited to appear, are examined on oath by the coroner, who must, in criminal cases at least, take down the evidence in writing. This is then read over to each witness, who signs it, and this forms his deposition. At the end of each case the coroner sums up, and the jury return their verdict or inquisition, either unanimously or by a majority.

If this charges any person with murder or manslaughter, he is committed by the coroner to prison to await trial, or, if not present, the coroner may issue a warrant for his arrest.

A chemical analysis of the contents of the stomach, etc., in suspected cases of poisoning is usually done by a special analyst named by the coroner. If any witness disobeys the summons to attend the inquest, he renders himself liable to a fine not exceeding £2 2s, but in addition the coroner may commit him to prison for contempt of court. In criminal cases the witnesses are bound over to appear at the assizes to give evidence there. The coroner may give an order for the exhumation of a body if he thinks the evidence warrants a post-mortem examination.

Coroners' inquests are held in all cases of sudden or violent death, where the cause of death is not clear; in cases of assault, where death has taken place immediately or some time afterwards; in cases of homicide or suicide; where the medical attendant refuses to give a certificate of death; where the attendants on the deceased have been culpably negligent; or in certain cases of uncertified deaths.

The medical witness should be very careful in giving evidence before a coroner. Even though the inquest be held in a coach-house or barn, yet it has to be remembered it is a court of law. If the case goes on for trial before a superior court, your deposition made to the coroner forms the basis of your examination. Any misstatements or discrepancies in your evidence will be carefully inquired into, and you will make a bad impression on judge and jury if you modify, retract, or explain away your evidence as given to the coroner. You had your opportunity of making any amendments on your evidence when the coroner read over to you your deposition before you signed it as true.

By the Licensing Act of 1902, an inquest may not be held in any premises licensed for the sale of intoxicating liquor if other suitable premises have been provided.

The duties of the coroner are based partly on Common Law, and are also defined by statute, principally by the Coroners Act of 1887 (50 and 51 Vict. c. 71). They have been modified, however, by subsequent Acts—e.g., the Act of 1892, the Coroners (Emergency Provisions) Act, 1917, and the Juries Act of 1918.

The fee payable to a medical witness for giving evidence at an inquest is one guinea, with an extra guinea for making a post-mortem examination and report (in the metropolitan area these fees are doubled). The coroner must sign the order authorizing the payment, and should an inquest be adjourned to a later day, no further fee is payable. If the deceased died in a hospital, infirmary, or lunatic asylum, the medical witness is not paid any fee. Should a medical witness neglect to make the post-mortem examination after receiving the order to do so, he is liable to a fine of £5.

In Scotland the Procurator Fiscal fulfils many of the duties of the coroner, but he cannot hold a public inquiry. He interrogates the witnesses privately, and these questions with the answers form the precognition. More serious cases are dealt with by the Sheriff of each county, and capital charges must be dealt with by the High Court of Justiciary. In Scotland the verdicts of the jury may be 'guilty,' 'not guilty,' or 'not proven.'

2. The Magistrate's Court or Petty Sessions is also a court of preliminary inquiry. The prisoner may be dealt with summarily, as, for example, in minor assault cases, or, if the case is of sufficient gravity, and the evidence justifies such a course, may be committed for trial. The fee for a medical witness who resides within three miles of the court is ten shillings and sixpence; if at a greater distance, one guinea.

In the Metropolis the prisoner in the first instance is brought before a magistrate, technically known as the 'beak,' who, in addition to being a person of great acumen, is a stipendiary, and thus occupies a superior position to the ordinary 'J.P.,' who is one of the great unpaid. In the City of London is the Mansion House Justice-Room, presided over by the Lord Mayor or one of the Aldermen. The prisoner may ultimately be sent for trial to the Central Criminal Court, known as the Old Bailey, or elsewhere.

3. Quarter Sessions.—These are held every quarter by Justices of the Peace. All cases can be tried before the sessions except felonies or cases which involve difficult legal questions. In London this court is known as the Central Criminal Court, and it also acts as the Assize Court. In Borough Sessions a barrister known as the Recorder is appointed as sole judge.

4. The Assizes deal with both criminal and civil cases. There is the Crown Court, where criminal cases are tried, and there is the Civil Court, where civil cases are heard. Before a case sent up by a lower court can be tried by the judge and petty jury, it is investigated by the grand jury, which is composed of superior individuals. If they find a 'true bill,' the case goes on; but if they 'throw it out,' the accused is at liberty to take his departure. At the Court of Assize the prisoner is tried by a jury of twelve. In bringing in the verdict the jury must be unanimous. If they cannot agree, the case must be retried before a new jury. At the Assize Court the medical witness gets a guinea a day, with two shillings extra to pay for his bed and board for every night he is away from home, with his second-class railway fare, if there is a second class on the railway by which he travels. If there is no railway, and he has to walk, he is entitled to threepence a mile for refreshments both ways.

5. Court of Criminal Appeal.—This was established in 1908, and consists of three judges. A right of appeal may be based (1) solely on a question of law; (2) on certificate from the judge who tried the prisoner; (3) on mitigation of sentence.

Speaking generally, in the Superior Courts the fees which may be claimed by medical men called on to give evidence are a guinea a day if resident in the town in which the case is tried, and from two to three guineas a day if resident at a distance from the place of trial, this to include everything except travelling expenses. The medical witness also receives a reasonable allowance for hotel and travelling expenses.

If a witness is summoned to appear before two courts at the same time, he must obey the summons of the higher court. Criminal cases take precedence of civil.

A medical man has no right to claim privilege as an excuse for not divulging professional secrets in a court of law, and the less he talks about professional etiquette the better. Still, in a civil case, if he were to make an emphatic protest, the matter in all probability would not be pressed. In a criminal case he would promptly be reminded of the nature of his oath.

A medical man may be required to furnish a formal written report. It may be the history of a fatal illness or the result of a post-mortem examination. These reports must be drawn up very carefully, and no technical terms should be employed.

No witness on being sworn can be compelled to 'kiss the book.' The Oaths Act (51 and 52 Vict., c. 46, § 5) declares, without any qualification, that 'if any person to whom an oath is administered desires to swear with uplifted hand, in the form and manner in which an oath is usually administered in Scotland, he shall be permitted to do so, and the oath shall be administered to him in such form and manner without further question.' The witness takes the oath standing, with the bare right hand uplifted above the head, the formula being: 'I swear by Almighty God that I will speak the truth, the whole truth, and nothing but the truth.' The presiding judge should say the words, and the witness should repeat them after him. There is no kissing of the book, and the words 'So help me, God,' which occur in the English form, are not employed. It will be noted that the Scotch form constitutes an oath, and is not an affirmation. The judge has no right to ask if you object on religious grounds, or to put any question. He is bound by the provisions of the Act, and the enactment applies not only to all forms of the witness oath, whether in civil or criminal courts, or before coroners, but to every oath which may be lawfully administered either in Great Britain or Ireland.

A witness engaged to give expert evidence should demand his fee before going into court, or, at all events, before being sworn.

With regard to notes, these should be made at the time, on the spot, and may be used by the witness in court as a refresher to the memory, though not altogether to supply its place. All evidence is made up of testimony, but all testimony is not evidence. The witness must not introduce hearsay testimony. In one case only is hearsay evidence admissible, and that is in the case of a dying declaration. This is a statement made by a dying person as to how his injuries were inflicted. These declarations are accepted because the law presumes that a dying man is anxious to speak the truth. But the person must believe that he is actually on the point of death, with absolutely no hope of recovery. A statement was rejected because the dying person, in using the expression 'I have no hope of recovery,' requested that the words 'at present' should be added. If after making the statement the patient were to say, 'I hope now I shall get better,' it would invalidate the declaration. To make the declaration admissible as evidence, death must ensue. If possible, a magistrate should take the dying declaration; but if he is not available, the medical man, without any suggestions or comments of his own, should write down the statements made by the dying person, and see them signed and witnessed. It must be made clear to the court that at the time of making his statement the witness was under the full conviction of approaching or impending death.

III

PERSONAL IDENTITY

It is but seldom that medical evidence is required with regard to the identification of the living, though it may sometimes be so, as in the celebrated Tichborne case. The medical man may in such cases be consulted as to family resemblance, marks on the body, nævi materni, scars and tattoo marks, or with regard to the organs of generation in cases of doubtful sex. Tattoo marks may disappear during life; the brighter colours, as vermilion, as a rule, more readily than those made with carbon, as Indian ink; after death the colouring-matter may be found in the proximal glands. If the tattooing is superficial (merely underneath the cuticle) the marks may possibly be removed by acetic acid or cantharides, or even by picking out the colouring-matter with a fine needle. With regard to scars and their permanence, it will be remembered that scars occasioned by actual loss of substance, or by wounds healed by granulation, never disappear. The scars of leech-bites, lancet-wounds, or cupping instruments, may disappear after a lapse of time. It is difficult, if not impossible, to give any certain or positive opinion as to the age of a scar; recent scars are pink in colour; old scars are white and glistening. The cicatrix resulting from a wound depends upon its situation. Of incised wounds an elliptical cicatrix is typical, linear being chiefly found between the fingers and toes. By way of disguise the hair may be dyed black with lead acetate or nitrate of silver; detected by allowing the hair to grow, or by steeping some of it in dilute nitric acid, and testing with iodide of potassium for lead, and hydrochloric acid for silver. The hair may be bleached with chlorine or peroxide of hydrogen, detected by letting the hair grow and by its unnatural feeling and the irregularity of the bleaching.

Finger-print impressions are the most trustworthy of all means of identification. Such a print is obtained by rubbing the pulp of the finger in lampblack, and then impressing it on a glazed card. The impression reveals the fine lines which exist at the tips of the fingers. The arrangement of these lines is special to each person, and cannot be changed. Hence this method is employed by the police in the identification of prisoners.

In the determination of cases of doubtful sex in the living, the following points should be noticed: the size of the penis or clitoris, and whether perforate or not, the form of the prepuce, the presence or absence of nymphæ and of testicles or ovaries. Openings must be carefully sounded as to their communication with bladder or uterus. After puberty, inquiry should be made as to menstrual or vicarious discharges, the general development of the body, the growth of hair, the tone of voice, and the behaviour of the individual towards either sex.

With regard to the identification of the dead in cases of death by accident or violence, the medical man's assistance may be called. The sex of the skeleton, if that only be found, may be judged from the bones of the female generally being smaller and more slender than those of the male, by the female thorax being deeper, the costal cartilages longer, the ilia more expanded, the sacrum flatter and broader, the coccyx movable and turned back, the tuberosities of the ischia wider apart, the pubes shallow, and the whole pelvis shallower and with larger outlets. But of all these signs the only one of any real value is the roundness of the pubic arch in the female, as compared with the pointed arch in the male. Before puberty the sex cannot be determined from an examination of the bones.

Age may be calculated from the presence, nature and number of the erupted teeth; from the cartilages of the ribs, which gradually ossify as age advances; from the angle formed by the ramus of the lower jaw with its body (obtuse in infancy, a right angle in the adult, and again obtuse in the aged from loss of the teeth); and in the young from the condition of the epiphyses with regard to their attachment to their respective shafts.

To determine stature, the whole skeleton should be laid out and measured, 1-1/2 to 2 inches being allowed for the soft parts.

IV

EXAMINATION OF PERSONS FOUND DEAD

When a medical man is called to a case of sudden death, he should carefully note anything likely to throw any light on the cause of death. He should notice the place where the body was found, the position and attitude of the body, the soil or surface on which the body lies, the position of surrounding objects, and the condition of the clothes. He should also notice if there are any signs of a struggle having taken place, if the hands are clenched, if the face is distorted, if there has been foaming at the mouth, and if urine or fæces have been passed involuntarily. Urine may be drawn off with a catheter and tested for albumin and sugar.

If required to make a post-mortem examination, every cavity and important organ of the body must be carefully and minutely examined, the seat of injury being inspected first.

V

MODES OF SUDDEN DEATH

There are three modes in which death may occur: (1) Syncope; (2) asphyxia; (3) coma.

1. Syncope is death beginning at the heart—in other words, failure of circulation. It may arise from—(1) Anæmia, or deficiency of blood due to hæmorrhage, such as occurs in injuries, or from bleeding from the lungs, stomach, uterus, or other internal organs. (2) Asthenia, or failure of the heart's action, met with in starvation, in exhausting diseases, such as phthisis, cancer, pernicious anæmia,

and Bright's disease, and in some cases of poisoning—for example, aconite.

The symptoms of syncope are faintness, giddiness, pallor, slow, weak, and irregular pulse, sighing respiration, insensibility, dilated pupils, and convulsions.

Post mortem the heart is found empty and contracted. When, however, there is sudden stoppage of the heart, the right and left cavities contain blood in the normal quantities, and blood is found in the venæ cavæ and in the arterial trunks. There is no engorgement of either lungs or brain.

2. Asphyxia, or death beginning at the lungs, may be due to obstruction of the air-passages from foreign bodies in the larynx, drowning, suffocation, strangling, and hanging; from injury to the cervical cord; effusion into the pleuræ, with consequent pressure on the lungs; embolism of the pulmonary artery; and from spasmodic contraction of the thoracic and abdominal muscles in strychnine-poisoning.

The symptoms of this condition are fighting for breath, giddiness, relaxation of the sphincters, and convulsions.

Post mortem, cadaveric lividity is well marked, especially in nose, lips, ears, etc.; the right cavities of the heart and the venæ cavæ are found gorged with dark fluid blood. The pulmonary veins, the left cavities of the heart, and the aorta, are either empty or contain but little blood. The lungs are dark and engorged with blood, and the lining of the air-tubes is bright red in colour. Much bloody froth escapes on cutting into the lungs. Numerous small hæmorrhages (Tardieu's spots) are found on the surface and in the substance of the internal organs, as well as in the skin of the neck and face.

3. Coma, or death beginning at the brain, may arise from concussion; compression; cerebral pressure from hæmorrhage and other forms of apoplexy; blocking of a cerebral artery from embolism; dietetic and uræmic conditions; and from opium and other narcotic poisons.

The symptoms of this condition are stupor, loss of consciousness, and stertorous breathing.

The post-mortem signs are congestion of the substance of the brain and its membranes, with accumulation of the blood in the cavities of the heart, more on the right side than on the left.

It must be remembered that, owing to the interdependence of all the vital functions, there is no line of demarcation between the various modes of death. In all cases of sudden death think of angina pectoris and the rupture of an aneurism.

The following is a list of some of the commoner causes of sudden death:

(a) Instantaneously Sudden Death—

Syncope (by far the commonest cause).
Aortic incompetence.
Rupture of heart.
Rupture of a valve.
Rupture of aortic aneurism.
Embolism of coronary artery.
Angina pectoris.

(b) Less Sudden but Unexpected Death—

Cerebral hæmorrhage or embolism.
Mitral and tricuspid valvular lesions if the patient exerts himself.
Rupture of a gastric or duodenal ulcer; rupture of liver, spleen, or extra-uterine gestation, or abdominal aneurism.
Suffocation during an epileptic fit; vomited matter or other material drawn into the trachea or air-passages; croup.
Arterio-sclerosis may lead to thrombosis, embolism, or aneurism.
Poisoning, as by hydrocyanic acid, cyanide of potassium, inhalation of carbonic acid or coal gas, œdema of glottis following inhalation of ammonia.
Rapid onset of some acute specific disease, such as pneumonia or diphtheria; collapse from cholera.

Heat-stroke, lightning, shocks of electricity of high tension.
Mental or physical shock.
Exertion while the stomach is overloaded.
Diabetic coma; uræmia.

Status lymphaticus. This is a general hyperplastic condition of the lymphatic structures in the body, and is seen in enlargement of tonsils, thymus, spleen, as well as of Peyer's patches and mesenteric glands. It is a frequent cause of death during chloroform anæsthesia for slight operations in young people.

In addition, it may be as well to remember that death sometimes occurs suddenly in exophthalmic goitre, hypertrophy of the thymus, and in Addison's disease.

In some cases of sudden death nothing has been found post mortem, even when the autopsy has been made by skilled observers, and the brain and cord have been submitted to microscopical examination.

VI

SIGNS OF DEATH

(1) Cadaveric appearance; ashy white colour. (2) Cessation of the circulation and respiration, no sound being heard by the stethoscope. Cessation of the circulation may be determined by (a) placing a ligature round the base of a finger (Magnus' test); (b) injecting a solution of fluorescin (Icard's test); (c) looking through the web of the fingers at a bright light (diaphanous test); (d) the dulling of a steel needle when thrust into the living body; (e) the clear outline of the dead heart when viewed in the fluorescent screen. (3) The state of the eye; the tension is at once lost; iris insensible to light, fundus yellow in colour; cornea dull and sunken.

(4) The state of the skin; pale, livid, with loss of elasticity. (5) Extinction of muscular irritability. The above signs afford no means of determining how long life has been extinct. The following, however, do:

Cooling of the Body.—The average internal temperature of the body is from 98° to 100° F. The time taken in cooling is from fifteen to twenty hours, but it may be modified by the kind of death, the age of the person, the presence or absence of clothing on the body, the surrounding temperature, and the stillness or otherwise of the air about the body. Still, the body, other things being equal, may be said to be quite cold in about twelve hours.

Hypostasis or post-mortem staining is due to the settling down of the blood in the most dependent parts of the body while the body is cooling. It is a sure sign of death, and occurs in all forms of death, even in that due to hæmorrhage, although not so marked in degree. Post-mortem staining (cadaveric lividity) begins to appear in from eight to twelve hours after death, and its position on the body will help to determine the length of time the body has lain in the position in which it was found. The staining is of a dull red or slaty blue colour. It must be distinguished from ecchymosis the result of a bruise, by making an incision into the part; in the case of hypostasis a few small bloody points of divided arteries will be seen, in the case of ecchymosis the subcutaneous tissues are infiltrated with blood-clot. Internally, hypostasis must not be mistaken for congestion of the brain or lungs, or the results of inflammation of the intestines. If the intestine is pulled straight, inflammatory redness is continuous, hypostasis is disconnected. About the neck hypostasis must not be mistaken for the mark of a cord or other ligature. When the blood is of a bright red colour after death (as happens in poisoning by CO or HCN, or in death from cold), the hypostasis is bright red also.

Cadaveric Rigidity—Rigor Mortis.—For some time after death the muscles continue to contract under stimuli. When this irritability ceases—and it seldom exceeds two hours—rigidity and hardening sets in, and in all cases precedes putrefaction. It is caused by the coagulation of the muscle plasma. It commences in the muscles of the back of the neck and lower jaw, and then passes into the muscles of the face, front of the neck, chest, upper extremities, and lastly to the lower extremities.

It has been noticed in the new-born infant, as well as in the fœtus. It lasts from sixteen to twenty hours or more. In lingering diseases, after violent exertion, and in warm climates, it sets in quickly, and disappears in two or three hours; in those who are in perfect health and die from accident or asphyxia, it may not come on until from ten to twenty-four hours, and may last three or four days. After death from convulsions or strychnine-poisoning, the body may pass at once into rigor mortis. Rigor mortis must be distinguished from cadaveric spasm or the death clutch; in the former, articles in the hands are readily removable, in the latter this is not the case. In tetanic spasm the limbs when bent return to their former position; not so in rigor mortis.

Putrefaction appears in from one to three days after death, as a greenish-blue discoloration of the abdomen; in the drowned, over the head and face. This increases, becomes darker and more general, a strong putrefactive odour is developed, the thorax and abdomen become distended with gas, and the epidermis peels off. The muscles then become pulpy, and assume a dark greenish colour, the whole body at length becoming changed into a soft, semi-fluid mass. The organ first showing the putrefactive change is the trachea; that which resists putrefaction longest is the uterus. These putrefactive changes are modified by the fat or lean condition of the body, the temperature (putrefaction taking place more rapidly in summer than in winter), access of air, the period, place, mode of interment, age, etc. Bodies which remain in water putrefy more slowly than those in air.

Saponification.—In bodies which are very fat and have lain in water or moist soil for from one to three years this process takes place, the fat uniting with the ammonia given off by the decomposition to form adipocere. This consists of a margarate or stearate of ammonium with lime, oxide of iron, potash, certain fatty acids, and a yellowish odorous matter. It has a fatty, unctuous feel, is either pure white or pale yellow, with an odour of decayed cheese. Small portions of the body may show signs of this change in six weeks.

Post-Mortem Examination.—Never make an autopsy in criminal cases without a written order from the coroner or Procurator Fiscal. If authorized, however, first have the body identified, then photographed if it has not been identified. A medical man representing the accused may be present, but only by consent of the Crown authorities or of the Sheriff. Clothing should be examined for blood-stains, cuts, etc.

Examine external surface of body and take accurate measurements of wounds, marks, deformities, tattooings; note degree and distribution of post-mortem staining, rigidity, etc.

Examine brain by making incision from ear to ear across vertex, reflect scalp forwards and backwards, and saw off calvarium. Examine brain carefully externally and on section.

Examine organs of chest and abdomen through an incision made from symphysis menti to pubis, reflecting tissues from chest wall and cutting through costal cartilages.

In cases of suspected poisoning have several clean jars into which you place the stomach with contents, intestines with contents, piece of liver, kidney, spleen, etc., and seal each up carefully, attaching label with name of deceased, date, and contained organs, and transmit these personally to the analyst.

Exhumation.—A body which has been buried cannot be exhumed without an order from a coroner, fiscal, or from the Home Secretary. There is no legal limit in England as to when a body may be exhumed; in Scotland, however, if an interval of twenty years has elapsed, an accused person cannot be prosecuted (prescription of crime).

VII

DEATH FROM ANÆSTHETICS, ETC.

The coroner in England and Wales and Ireland must inquire into every case of death during the administration of an anæsthetic. The anæsthetist has to appear at the inquest, and must answer a long series of questions relative to the administration of the drug.

Before, therefore, giving an anæsthetic, and so as to furnish yourself with a proper defence in the event of death occurring, you ought to examine the heart, lungs, and kidneys of the patient to see if they are healthy. Should a fatal result follow, the anæsthetist will require to prove that it was necessary to give the anæsthetic, that the one employed was the most suitable, that the patient was in a fit state of health to have it administered, that it was given skilfully and in moderate amount, that he had the usual remedies at hand in case of failure of the heart or lungs, and that he employed every means in his power to resuscitate the patient.

The condition of the lungs is of more importance than the state of the heart.

The chloroformist ought always to use the best chloroform.

An anæsthetic should never be administered except in the presence of a third person. This applies especially to dentists who give gas to females.

Malpractice.—In every case where a medical man attends a patient, he must give him that amount of care, skill, knowledge, or judgment, that the law expects of him. If he does not, then the charge of malpractice may be brought against him. It is most frequently alleged in connection with surgical affections—e.g., overlooking a fracture or dislocation. Before a major operation is performed, it is well to get a written agreement.

VIII

PRESUMPTION OF DEATH; SURVIVORSHIP

Presumption of Death.—If a person be unheard of for seven years, the court may, on application by the nearest relative, presume death to have taken place. If, however, it can be shown that in all probability death had occurred in a certain accident or shipwreck, the decree may be made much earlier.

Presumption of Survivorship.—When two or more related persons perish in a common accident, it may be necessary, in order to decide questions of succession, to determine which of them died first. It is generally accepted that the stronger and more vigorous will survive longest.

IX

ASSAULT, MURDER, MANSLAUGHTER, ETC.

Assault.—This is an attempt or offer to do violence to another person; it is not necessary that actual injury has been done, but evil intention must be proved. When a corporal hurt has been sustained, then assault and battery has been committed. The assault may be aggravated by the use of weapons, etc.

Homicide may be justifiable, as in the case of judicial execution, or excusable, as in defence of one's family or property.

Felonious homicide is murder. This means that a human being has been killed by another maliciously and deliberately or with reckless disregard of consequences.

Manslaughter or Culpable Homicide (Scotland) is the unlawful killing of a human being without malice—as homicide after great provocation; signalman who allows a train to pass, and so collide with another in front.

X

WOUNDS AND MECHANICAL INJURIES

A wound may be defined as a 'breach of continuity in the structures of the body, whether external or internal, suddenly occasioned by mechanical violence.' The law does not define 'a wound,' but the true skin must be broken. Wounds are dangerous from shock, hæmorrhage, from the supervention of crysipelas or pyæmia, and from malum regimen on the part of the patient or surgeon. Is the wound dangerous to life? This question can only be answered by a full consideration of all the circumstances of the case; a guarded prognosis is wise in all cases.

Burns are caused by flames, highly heated solids, or very cold solids, as solid carbonic acid; scalds, by steam or hot fluids. Burns may cause death from shock, suffocation, œdema glottidis, inflammation of serous surfaces, bronchitis, pneumonia, duodenal ulcer, coma, or exhaustion. A burn of the skin inflicted during life is followed by a bleb containing serum; the edges of this blister are bright red, and the base, seen after removing the cuticle, is red and inflamed; if sustained after death, a bleb, if present, contains but little fluid, and there are no signs of vital reaction. There are six degrees of burns: (1) Superficial inflammation; (2) formation of vesicles; (3) destruction of superficial layer of skin; (4) destruction of cellular tissue; (5) deep parts charred; (6) carbonization of bones.

The larger the area of skin burnt, the more grave is the prognosis.

Burns of the abdomen and genital organs are especially dangerous. Young children are specially liable to die after burns.

XI

CONTUSED WOUNDS AND INJURIES UNACCOMPANIED BY SOLUTION OF CONTINUITY

If a blow be inflicted with a blunt instrument, there is produced a bruise, or ecchymosis, of which it is unnecessary here to describe the appearance and progress. A bruise may be distinguished from a post-mortem stain by the cuticle in the former often being abraded and raised. When an incision is made into the bruise, the whole of the subcutaneous tissues are found to be infiltrated with blood-clot, and there is no clear margin. In the case of a post-mortem stain the edges are sharply defined, not raised, and, on section, mere bloody points are seen which are the cut ends of the divided blood vessels.

XII

INCISED WOUNDS AND THOSE ACCOMPANIED BY SOLUTION OF CONTINUITY

These comprise incised, punctured, and lacerated wounds. In a recent incised wound inflicted during life there is copious hæmorrhage, the cellular tissue is filled with blood, the edges of the wound gape and are everted, and the cavity of the wound is filled with coagula.

Lacerated wounds combine the characters of incised and contused wounds. They are caused by falls, being ridden over, machinery crushes, bites, blows from blunt weapons, etc. The wounds heal by suppuration.

Punctured wounds come intermediate between incised and lacerated. They are greater in depth than in length, being caused by sword or rapier thrusts. They cause little hæmorrhage externally, but death may be due to internal hæmorrhage. They may be complicated by (1) the introduction of septic material adhering to the instrument; (2) the entrance of foreign bodies which lodge in the wound, not only carrying in septic matter, but acting as mechanical irritants; (3) injury to deeper parts, which may at the time be difficult to detect.

An apparently incised wound may be produced by a hard, blunt weapon over a bone—e.g., shin or cranium. It is often difficult to distinguish between a wound of the scalp inflicted with a knife and one made by a blow with a stick. A puncture with a sharp-edged, pointed knife leaves a fusiform or spindle-shaped wound. A wound from a blow with a stick might be of this character, or it might present a jagged, swollen appearance at the margin, with much contusion of the surrounding tissues. If the wound is seen soon after it is inflicted, examination with a lens may disclose irregularities of the margins, or little bridges of connective tissue or vessels running across the wound, and so be inconsistent with its production by a cutting instrument. Lacerated wounds as a rule bleed less freely than those which are incised. Symptoms of concussion would favour the theory of the injury having been inflicted by a heavy instrument. Again, it is often difficult to decide whether the injury which caused death was the result of a blow or a fall. A heavy blow with a stick may at once cause fatal effusion of blood, but this might equally result from fracture of the skull resulting from a fall. The wound should be carefully examined for foreign bodies, such as grit, dirt, or sand. The distinction between incised wounds inflicted during life and after death is found in the fact that a wound inflicted during life presents the appearances already described, whereas in a post-mortem incised wound only a small quantity of liquid venous blood is effused; the edges are close, yielding, inelastic; the blood is not effused into the cellular tissue, and there are no signs of vital reaction. The presence of inflammatory reaction or pus shows that the wound must have been inflicted some time before death, probably two or three days.

Self-inflicted wounds are made by the person himself in order to divert suspicion, or in order to bring accusation against another. Such wounds are always in front, not over vital organs, and superficial in character. Note the condition of the clothes in such cases.

XIII

GUNSHOT WOUNDS

These may be punctured, contused, or lacerated. Round balls make a larger opening than those which are conical. Small shot fired at a short distance make one large ragged opening; while at distances greater than 3 feet the shot scatter and there is no central opening. The Lee-Metford bullet is more destructive than the Mauser. The former is the larger, but the difference in size is not great. The Martini-Henry bullet weighs 480 grains, the Lee-Metford 215, and the Mauser 173. Speaking generally, a gunshot wound, unlike a punctured wound, becomes larger as it increases in depth; the aperture of entrance is round, clean, with inverted edges, and that of exit larger, less regular than that of entrance, and with everted edges.

In the case of high-velocity bullets from smooth-bore rifles, including the Mauser and Lee-Metford, the aperture of entry is small; the aperture of exit is slightly larger, and tends to be more slit-like. There is but little tendency to carry in portions of clothing or septic material, and the wound heals by first intention, if reasonable precautions be taken. The external cicatrices finally look very similar to those produced by bad acne pustules.

The contents of all gunshot wounds should be preserved, as they may be useful in evidence. A pocket revolver, as a rule, leaves the bullet in the body.

Wounds inflicted by firearms may be due to accident, homicide, or suicide. Blackening of the wound, singeing of the hair, scorching of the skin and clothing, show that the weapon was fired at close quarters, whilst blackening of the hand points to suicide. Even when the weapon is fired quite close there may be no blackening of the skin, and the hand is not always blackened in cases of suicide. Smokeless powder does not blacken the skin. Wounds on the back of the body are not usually self-inflicted, but a suicide may elect to blow off the back of his head. A wound in the back may be met with in a sportsman who indulges in the careless habit of dragging a loaded gun after him. If a revolver is found tightly grasped in the hand it is probably a case of suicide, whilst if it lies lightly in the hand it may be suicide or homicide. If no weapon is found near the body, it is not conclusive proof that it is not suicide, for it may have been thrown into a river or pond, or to some distance and picked up by a passer-by.

A bullet penetrating the skull even from a distance of 3,000 yards may act as an explosive, scattering the contents in all directions; but the bullet from a revolver will usually be found in the cranium.

The prognosis depends partly on the extent of the injury and the parts involved, but there is also risk from secondary hæmorrhage, and from such complications as pleurisy, pericarditis, and peritonitis. Death may result from shock, hæmorrhage, injury to brain or important nervous structures.

XIV

WOUNDS OF VARIOUS PARTS OF THE BODY

1. Of the Head.—Wounds of the scalp are likely to be followed by (1) erysipelatous inflammation; (2) inflammation of the tendinous structures, with or without suppuration. A severe blow on the vertex may cause fracture of the base of the skull. Injuries of the brain include concussion, compression, wounds, contusion, and inflammation. Concussion is a common effect of blows or violent shocks, and the symptoms follow immediately on the accident, death sometimes taking place without reaction. Compression may be caused by depressed bone or effused blood (rupture of middle meningeal artery) and serum. The symptoms may come on suddenly or gradually. Wounds of the brain present very great difficulties, and vary greatly in their effect, very slight wounds producing severe symptoms, and vice versâ. A person may receive an injury to the head, recover from the first effects, and then die with all the symptoms of compression from internal hæmorrhage. This is due to the fact that the primary syncope arrests the hæmorrhage, which returns during the subsequent reaction, or on the occurrence of any excitement. Inflammation of the meninges or brain may follow injuries, not only to the brain itself, but to the scalp and adjacent parts, as the orbit and ear. Inflammation does not usually come on at once, but after variable periods.

2. Injuries to the Spinal Cord may be due to concussion, compression (fracture-dislocation), or wounds. That the wound has penetrated the meninges is shown by the escape of cerebro-spinal fluid. The cord and nerves may be injured (1) by the puncture; (2) by extravasation of blood and the formation of a clot; and (3) by subsequent septic inflammation. Division or complete compression of the cord at or above the level of the fourth cervical vertebra is immediately fatal (as happens in judicial hanging). When the injury is below the fourth, the diaphragm continues forcibly in action, but the lungs are imperfectly expanded, and life will not be maintained for more than a day or two. When the injury is in the dorsal region, there is paralysis of the legs and of the sphincters of the bladder and rectum, but power is retained in the arms and the upper intercostal muscles act, the extent of paralysis depending on the level of the lesion. In injuries to the lumbar region the legs may be partly paralysed, and the rectal and bladder sphincters may be involved.

Railway spine, or traumatic neurasthenia, may be set up by concussion of the cord as a result of blows or falls. Passengers after railway accidents, or miners, often suffer from this affection.

3. Of the Face.—These produce great disfigurement and inconvenience, and there is a risk of injury to the brain. The seventh nerve may be involved, giving rise to facial paralysis. Punctured wounds of the orbit are especially dangerous. Wounds apparently confined to the external parts often conceal deep-seated mischief.

4. Of the Eye.—The iris may be injured by sharp blows, as from the cork of a soda-water bottle. It is usually followed by hæmorrhage into the anterior chamber, and there may be separation of the iris from its ciliary border. Wounds at the edge of the cornea are often followed by prolapse of the iris. Acute traumatic iritis or iridocyclitis may supervene four or five days after the injury. The lens is frequently wounded in addition to the cornea and iris. In dislocation of the lens into the anterior chamber as the result of a blow, the lens appears like a large drop of oil lying at the back of the cornea, the margin exhibiting a brilliant yellow reflex. Partial dislocations of the lens as the result of severe blows generally terminate in cataract.

5. Of the Throat.—Very frequently inflicted by suicides. Division of the carotid artery is fatal, and of the internal jugular vein very dangerous on account of entrance of air. Wounds of the larynx and trachea are not necessarily or immediately dangerous, but septic pneumonia is very apt to follow. Wounds of the throat inflicted by suicides are commonly situated at the upper part, involving the hyoid bone and the thyroid and cricoid cartilages. The larynx is opened, but the large vessels often escape. In most suicidal wounds of the throat the direction is from left to right, the incision being slightly inclined from above downwards. At the termination of a suicidal cut-throat the skin is the last structure divided, the wound being shallower as it reaches its termination; the wounds often show parallelism. The weapon is often firmly grasped in the hand. Inquiry should be made as to whether the patient is right or left handed, or ambidextrous.

Homicidal cut throat is usually very severe and situated low down in the neck or far to the side.

6. Of the Chest.—Incised wounds of the walls are not of necessity dangerous; but severe blows, by causing fracture of the bones and internal injuries, are often fatal. The symptoms of penetrating wounds of the chest are—(1) The passage of blood and air through the wound; (2) hæmoptysis; (3) pneumothorax; and (4) protrusion of the lung forming a tumour covered with pleura. Fracture of the ribs may be due to direct violence, as from a blow, when the ends are driven inwards, or to indirect violence, as from a squeeze in a crowd, when the ends are driven outwards.

7. Of the Lungs.—These usually cause hæmorrhage, and are frequently followed by pleurisy, either dry or with effusion, and by pneumonia.

8. Of the Heart.—Penetrating wounds are fatal from hæmorrhage, of the base more speedily than of the apex; but life may be prolonged for some time even after a severe wound to the heart. Injury to the right ventricle is the most fatal injury and the most frequent. Rupture from disease usually occurs in the left ventricle; rupture from a crush is usually towards the base and on the right side.

9. Of the Aorta and Pulmonary Artery.—Fatal.

10. Of the Diaphragm.—Generally fatal, owing to the severe injury of the other abdominal organs. If the diaphragm be ruptured, hernia of the organs may result.

11. Of the Abdomen.—Of the walls, may be dangerous from division of the epigastric artery; ventral hernia may follow, internal hæmorrhage, etc. Blows on the abdomen are prone to cause death from cardiac inhibition.

12. Of the Liver.—May divide the large vessels. Venous blood flows profusely from a punctured wound of the liver. Wounds of the gall-bladder cause effusion of bile and peritoneal inflammation. Laceration of the liver may result from external violence without leaving any outward sign of the injury; it is commonly fatal. There is rapid and acute anæmia from the pouring out of blood into the abdominal cavity. This may also occur with injuries of other organs in the abdomen.

13. Of the Spleen.—Fatal hæmorrhage may result from penetrating wounds or from rupture due to kicks, blows, crushes, especially if the spleen be enlarged.

14. Of the Stomach.—May be fatal from shock, from hæmorrhage, from extravasation of contents, or from inflammation. The danger is materially lessened by prompt surgical intervention.

15. Of the Intestines.—May be fatal in the same way as those of the stomach. More dangerous in the small than in the large intestines.

16. Of the Kidneys.—May prove fatal from hæmorrhage, extravasation of urine, or inflammation.

17. Of the Bladder.—Dangerous from extravasation of urine. In fracture of the pelvis the bladder is often injured, and extraperitoneal infiltration of urine occurs, with frequently a fatal issue.

18. Of Genital Organs.—Incised wounds of penis may produce fatal hæmorrhage. Removal of testicles may prove fatal from shock to nervous system. Wounds of the spermatic cord may be dangerous from hæmorrhage. Wounds to the vulva are dangerous, owing to hæmorrhage from the large plexus of veins without valves.

XV

DETECTION OF BLOOD-STAINS, ETC.

Stains may require detection on clothing, on cutting instruments, on floors and furniture, etc. The following are the distinctive characters of blood-stains:

(a) Ocular Inspection.—Blood-stains on dark-coloured materials, which in daylight might be easily overlooked, may be readily detected by the use of artificial light, as that of a candle, brought near the cloth. Blood-spots when recent are of a bright red colour if arterial, of a purple hue if venous, the latter becoming brighter on exposure to the air. After a few hours blood-stains assume a reddish-brown or chocolate tint, which they maintain for years. This change is due to the conversion of hæmoglobin into methæmoglobin, and finally into hæmatin. The change of colour in warm weather usually occurs in less than twenty-four hours. The colour is determined, not entirely by the age of the stain, but is influenced by the presence or absence of impurities in the air, such as the vapours of sulphurous, sulphuric, and hydrochloric acids. If recent, a jelly-like material may be seen by the aid of a magnifying-glass lying between the fibres. If old, a cinnabar-red streak is seen on drawing a needle across the stain.

(b) Microscopic Demonstration.—With the aid of the microscope, blood may be detected by the presence of the characteristic blood-corpuscles. The human blood-corpuscle is a non-nucleated, biconcave disc, having a diameter of about 1/3500 of an inch. All mammalian red corpuscles have the same shape, except those of the camel, which are oval. The corpuscles of birds, fishes, reptiles, and amphibians, are oval and nucleated. The corpuscles of most mammals are smaller than those of man, but the size of a corpuscle is affected by various circumstances, such as drying or moisture, so that the medical witness is rarely justified in going farther than stating whether the stain is that of the blood of a mammal or not. Unfortunately, the corpuscles are usually so dried that little information regarding their size can be given.

(c) Action of Water.—Water has a solvent action on blood, fresh stains rapidly dissolving when the material on which they occur is placed in cold distilled water, forming a bright red solution. The hæmatin of old stains dissolves very slowly, so employ a weak solution of ammonia, and this will give a solution of alkaline hæmatin. Rust is not soluble in water.

(d) Action of Heat.—Blood-stains on knives may be removed by heating the metal, when the blood will peel off, at once distinguishing it from rust. Should the blood-stain on the metal be long exposed to the air, rust may be mixed with the blood, when the test will fail. The solution obtained in water is coagulated by heat, the colour entirely destroyed, and a flocculent muddy-brown precipitate formed.

(e) Action of Caustic Potash.—The solution of blood obtained in water is boiled, when a coagulum is formed soluble in hot caustic potash, the solution formed being greenish by transmitted and red by reflected light.

(f) Action of Nitric Acid.—Nitric acid added to a watery solution produces a whitish-grey precipitate.

(g) Action of Guaiacum.—Tincture of guaiacum produces in the watery solution a reddish-white precipitate of the resin, but on addition of an aqueous solution of peroxide of hydrogen, or of an ethereal solution of the same substance (known as ozonic ether), a blue or bluish-green colour is developed. This test is delicate, and succeeds best in dilute solutions. It is not absolutely indicative of the presence of blood, for tincture of guaiacum is coloured blue by milk, saliva, and pus.

(h) Hæmin Crystals (Teichman's Crystals).—These are produced by heating a drop of blood, or a watery solution of it, with a minute crystal of sodium chloride on a glass slide and evaporating to dryness. A cover-glass is placed over this, and a drop of glacial acetic acid allowed to run in. It is again heated until bubbles appear. Crystals of hæmin may now be detected by the microscope. They are dark brown or yellow rhombic prisms.

An improvement on this test is the use of formic acid alone; on slowly evaporating it, numerous very small dark crystals are visible if hæmoglobin has been present (Whitney's test).

(i) Spectroscopic Appearances.—If a solution of a recent stain be examined by the spectroscope, we get two absorption bands situated between the lines D and E, the one nearer E being doubly as broad as the other. These bands indicate oxyhæmoglobin.

If we now add a little ammonium sulphide to this solution, we get the spectrum of reduced hæmoglobin, which is a single broad absorption band situated in the interval between the preceding oxyhæmoglobin bands. By shaking the solution, oxyhæmoglobin is again reproduced, and gives its special absorption bands.

If ammonia be added to the original solution, alkaline hæmatin is produced, or if acetic acid be chosen, acid hæmatin is produced, and each gives its appropriate absorption bands.

Methæmoglobin is formed in stains which have been exposed to the air for a few days, and hæmatin is found in old stains. Hæmochromogen gives a very characteristic spectrum, and is obtained by reducing alkaline hæmatin by ammonium sulphide. Carbon monoxide hæmoglobin gives a spectrum which resembles that of oxyhæmoglobin, but it is not reduced by ammonium sulphide.

(j) Precipitin Test.—This allows us to tell whether the blood is from a human being or not. A specific serum must be obtained from a rabbit which is sensitized as follows: 10 c.c. of human blood is injected into its peritoneal cavity at intervals, until from three to five injections have been given. The serum of this animal's blood will then give a white precipitate only when brought into contact with dilute solutions of human blood, but with the blood of no other animal. This is known also as the 'biologic,' or Uhlenhuth's test.

Rust Stains.—These are yellowish-red in colour, and do not stiffen the cloth. The iron may be dissolved by placing the stain in a dilute solution of hydrochloric acid, when, on adding ferrocyanide of potassium, Prussian blue is produced.

Fruit Stains are seldom so dark as blood-stains. Solutions of these do not change colour or coagulate on boiling; ammonia changes the colour to blue or green; acid brightens the original colour, while chlorine bleaches it.

Hairs.—Human hairs must be identified and distinguished from those of the lower mammals. If the hair has been pulled out from the root, the microscope will show that the bulbous root has a concave surface which fitted over the hair papilla, or that the root is encased in a fatty sheath.

Fibres of Clothing.—Microscopically, wool fibres are coarse, curly, and striated transversely; cotton fibres appear as flattened bands twisted into spirals; linen fibres are round, jointed at frequent intervals, with small root-like filaments; silk fibres are solid, continuous, and highly glistening.

XVI

DEATH BY SUFFOCATION

Signs and Symptoms.—There are usually three stages:

1. Exaggerated respiratory activity; air hunger; anxiety; congested appearance of face; ringing in ears.

2. Loss of consciousness; convulsions; relaxation of sphincters.

3. Respirations feeble and gasping, and soon cease; convulsions of stretching character; heart continues to beat for three to four minutes after breathing ceases.

Post-Mortem Appearances—External.—Cadaveric lividity well marked; nose, lips, ears, finger-tips almost black in colour; appearance may be placid or, if asphyxia has been sudden, the tongue may be protruded and eyeballs prominent, with much bloody mucus escaping from mouth and nose.

Internal.—The blood is dark and remains fluid; great engorgement of venous system, right side of heart, great veins of thorax and abdomen, liver, spleen, etc. Lungs dark purple in colour; much bloody froth escapes on squeezing them; mucous lining of trachea and bronchi congested and bright red in colour; air-cells distended or ruptured; many small hæmorrhages on surface of lungs and other organs, as well as in their substance (Tardieu's spots), due to rupture of venous capillaries from increased vascular pressure.

XVII

DEATH BY HANGING

In hanging, death occurs by asphyxia, as in drowning. Sensibility is soon lost, and death takes place in four or five minutes. The eyes in some cases are brilliant and staring, tongue swollen and livid, blood or bloody froth is found about the mouth and nostrils, and the hands are clenched. In other cases the countenance is placid, with an almost entire absence of the signs just given. The mark on the neck, which may be more or less interrupted by the beard, shows the course of the cord, which in hanging is obliquely round the neck following the line of the jaw, but straight round in strangulation. In judicial hanging, death is not due to asphyxiation, but, owing to the long drop, the cervical vertebræ are dislocated, and the spinal cord injured so high up that almost instant death takes place. On dissection the muscles and ligaments of the windpipe may be found stretched, bruised, or torn, and the inner coats of the carotid arteries are sometimes found divided. In ordinary suicidal hanging there may be entire absence of injury to the soft parts about the neck, the length of the drop modifying these appearances. The mark of the cord is not a sign of hanging, is a purely cadaveric phenomenon, and may be produced some hours after death.

XVIII

DEATH BY STRANGULATION

This differs from hanging in that the body is not suspended. It may be effected by a ligature round the neck, or by direct pressure on the windpipe with the hand, in which case death is said to be caused by throttling. Strangulation is frequently suicidal, but may be accidental. When homicidal, much injury is done to the neck, owing to the force with which the ligature is drawn. In throttling, the marks of the finger-nails are found on the neck.

XIX

DEATH BY DROWNING

Death by drowning occurs when breathing is arrested by watery or semi-fluid substances—blood, urine, etc. The fluid acts mechanically by entering the air-cells of the lung and preventing the due oxidation of the blood. The post-mortem appearances include those usually present in death by asphyxia, together with the following, peculiar to death by drowning: Excoriations of the fingers, with sand or mud under the nails; fragments of plants grasped in the hand; water in the stomach (this is a vital act, and shows that the person fell into the water alive); fine froth at the mouth and nostrils; cutis anserina; retraction of penis and scrotum. On post-mortem examination, the lungs are found to be increased in size ('ballooned'); on section, froth, water mud, sand, in air-tubes. The presence of this fine (often blood-stained) froth is the most characteristic sign of drowning. Froth like that of soap-suds in the trachea is an indication of a vital act, and must not be mistaken for the tenacious mucus of bronchitis. The presence of vomited matters in the trachea and bronchi is a valuable sign of drowning. The blood collects in the venous system, and is dark and fluid. Tardieu's spots are not so frequently met with in cases of drowning as in other forms of asphyxia. The other signs of death by asphyxia are present. Wounds may be present on the body, due to falling on stakes, injuries from passing vessels, etc.

The methods of performing artificial respiration in the case of the apparently drowned are the following (the best and most easily performed is Schäfer's prone pressure method):

1. Schäfer's.—Place the patient on his face, with a folded coat under the lower part of the chest. Unfasten the collar and neckband. Go to work at once. Kneel over him athwart or on one side facing his head. Place your hands flat over the lower part of his back, and make pressure on his ribs on both sides, and throw the weight of your body on to them so as to squeeze out the air from his chest. Get back into position at once, but leave your hands as they were. Do this every five seconds, and get someone to time you with a watch. Keep this going for half an hour, and when you are tired get someone to relieve you.

Other people may apply hot flannels to the limbs and hot water to the feet. Hypodermic injections of 1/50 grain of atropine, suprarenal or pituitary extracts, may be found useful.

2. Silvester's..—In this method the capacity of the chest is increased by raising the arms above the head, holding them by the elbows, and thus dragging upon and elevating the ribs, the chest being emptied by lowering the arms against the sides of the chest and exerting lateral pressure on the thorax. The patient is in the supine position—but first the water must have been drained from the mouth and nose by keeping the body in the prone position. The tongue must be kept forward by transfixing with a pin.

3. Marshall Hall's.—This consists in placing the patient in the prone position, with a folded coat under the chest, and rolling the body alternately into the lateral and prone positions.

4. Howard's.—This consists in emptying the thorax by forcibly compressing the lower part of the chest; on relaxing the pressure the chest again fills with air. Here the patient is placed in the supine position.

The objections to the supine position are that the tongue falls back, and not only blocks the entrance of air, but prevents the escape of water, mucus, and froth from the air-passages.

5. *Laborde's Method.*—This consists in holding the tongue by means of a handkerchief, and rhythmically drawing it out fully at the rate of fifteen times per minute. This excites the respiratory centre, and this method may be employed along with any of the other methods.

XX

DEATH FROM STARVATION

The post-mortem appearances in death from starvation are as follows: There is marked general emaciation; the skin is dry, shrivelled, and covered with a brown, bad-smelling excretion; the muscles soft, atrophied, and free from fat; the liver is small, but the gall-bladder is distended with bile. The heart, lungs, and internal organs are shrivelled and bloodless. The stomach is sometimes quite healthy; in other cases it may be collapsed, empty, and ulcerated. The intestines are also contracted, empty, and translucent.

In the absence of any disease productive of extreme emaciation (e.g., tuberculosis, stricture of œsophagus, diabetes, Addison's disease), such a state of body will furnish a strong presumption of death by starvation.

In the case of children there is not always absolute deprivation of food, but what is supplied is insufficient in quantity or of improper quality. The defence commonly set up is that the child died either of marasmus or of tuberculosis.

In cases where it is alleged that a child has been starved and ill-used, one must examine the body for signs of neglect—e.g., dirtiness of skin and hair, presence of vermin, bruises or skin eruptions. Compare its weight with a normal child of the same age and sex. If the disproportion be great and signs of neglect present, then the probability is great (provided there be no actual disease present) that the child has been starved.

XXI

DEATH FROM LIGHTNING AND ELECTRICITY

The signs of death from lightning vary greatly. In some cases there are no signs; in others the body may be most curiously marked. Wounds of various characters—contused, lacerated, and punctured—may be produced. There may be burns, vesications, and ecchymoses; arborescent markings are not uncommon. The hair may be singed or burnt and the clothing damaged. Rigor mortis is very rapid in its onset and transient. Post mortem there are no characteristic signs, but the blood may be dark in colour and fluid. The presence or absence of a storm may assist the diagnosis.

Injuries by electrical currents of high pressure are not uncommon; speaking generally, 1,000 to 2,000 volts will kill. In America, where electricity is adopted as the official means of destroying criminals, 1,500 volts is regarded as the lethal dose, but there are many instances of persons having been exposed to higher voltages without bad effects. The alternating current is supposed to be more fatal than the continuous. Much depends on whether the contact is good (perspiring hands or damp clothes). Death has been attributed in these cases to respiratory arrest or sudden cessation of the heart's action. The best treatment is artificial respiration, but the inhalation of nitrite of amyl may prove useful. Rescuers must be careful that they, also, do not receive a shock. The patient should be handled with india-rubber gloves or through a blanket thrown over him.

XXII

DEATH FROM COLD OR HEAT

Cold.—The weak, aged, or infants, readily succumb to low temperatures. The symptoms are increasing lassitude, drowsiness, coma, with sometimes illusions of sight. Post mortem, bright red patches are found on the skin surface, and the blood remains fluid for long.

Heat.—Death may result from syncope, the result of exposure to great heat.

Sunstroke.—The person loses consciousness and falls down insensible; the body temperature may be 112° F., the pulse is full, and a peculiar pungent odour is given off from the skin. Coma, convulsions with (rarely) delirium, may precede death. Treatment consists in lowering the body temperature by application of cold cloths, stimulants, strychnine or digitalin hypodermically.

XXIII

PREGNANCY

The signs of the existence of pregnancy are of two kinds, uncertain and certain, or maternal and fœtal. Amongst the former class are included—Cessation of menstruation (which may occur without pregnancy); morning vomiting; salivation; enlargement of the breasts and of the abdomen; quickening. It must be borne in mind that every woman with a big abdomen is not necessarily pregnant. The tests which afford conclusive evidence of the existence of a fœtus in the uterus are—Ballottement, the uterine souffle, intermittent uterine contractions, fœtal movements, and, above all, the pulsation of the fœtal heart. The uterine souffle is synchronous with the maternal pulse; the fœtal heart is not, being about 120 beats per minute.

Evidence of pregnancy may also be afforded by the discharge from the uterus of an early ovum, of moles, hydatids, etc. Disease of the uterus and ovarian dropsy may be mistaken for pregnancy. Careful examination is necessary to determine the nature of the condition present. Pregnancy may be pleaded in bar of immediate capital punishment, in which case the woman must be shown to be 'quick with child.' A woman may also plead pregnancy to delay her trial in Scotland, and both in England and Scotland, in civil cases, to produce a successor to estates, to increase damages for seduction, in compensation cases where a husband has been killed, to obtain increased damages, etc. A woman may become pregnant within a month of her last delivery.

In cases of rape and suspected pregnancy, it must be borne in mind that a medical man who examines a woman under any circumstances against her will renders himself liable to heavy damages, and that the law will not support him in so doing. If, on being requested to permit an examination, the woman refuse, such refusal may go against her, but of this she is the best judge. The duty of the medical man ends on making the suggestion.

XXIV

DELIVERY

The signs of recent delivery are as follows: The face is pale, with dark circles round the eyes; the pulse quickened; the skin soft, warm, and covered with a peculiar sweat; the breasts full, tense, and knotty; the abdomen distended, its integuments relaxed, with irregular light pink streaks on the lower part. The labia and vagina show signs of distension and injury. For the first three or four days there is a discharge from the uterus more or less sanguineous in character, consisting of blood, mucus, epithelium, and shreds of membrane. During the next four or five days it becomes of a dirty green colour, and in a few days more of a yellowish, milky, mucous character, continuing for two to three weeks. The change in character of the lochial discharge is due to the quantity of blood decreasing and its place being taken by fatty granules and leucocytes. The os uteri is soft, patulous, and its edges are torn. The uterus may be felt for two or three hours above the pubis as a hard round ball, regaining its normal size in about eight weeks after delivery. Most of these signs disappear about the tenth day, after which it becomes impossible to fix the date of delivery.

In the dead the external parts have the same appearance as given above. The uterus will vary in appearance according to the time elapsed since delivery. If death occurred immediately after delivery, the uterus will be wide open, about 9 or 10 inches long, with clots of blood inside, and the inner surface lined by decidua.

The signs of a previous delivery consist in silvery streaks in the skin of the abdomen, which, however, may be due to distension from other causes; similar marks on the breast; circular and jagged condition of the os uteri (the virgin os being oval and smooth); marks of rupture of the perineum or fourchette; absence of the vaginal rugæ; dark-coloured areola round the nipples, etc. The difference between the virgin corpus luteum and that of recent pregnancy is not so marked as to justify a confident use of it for medico-legal purposes.

XXV

FŒTICIDE, OR CRIMINAL ABORTION

This consists in giving to any woman, or causing to be taken by her, with intent to procure her miscarriage, any poison or other noxious thing, or using for the same purpose any instruments or other means whatsoever. It is a felony to procure or attempt to procure the miscarriage of a woman, whether she be pregnant or not, and it is a felony for the woman, if pregnant, to attempt to procure her own miscarriage. It is a misdemeanour for any person or persons to procure drugs or instruments for a like purpose. It is not necessary that the woman be quick with child. The offence is the intent to procure the miscarriage of any woman, whether she be or be not with child. When from any causes it is necessary to procure abortion, a medical man should do so only after consultation with a brother practitioner. Even in these cases there is no exemption legally. Any medical man who gives even the most harmless medicine where he suspects the possibility of pregnancy may render himself liable to grave suspicion should the woman abort.

In medicine, an abortion is said to occur when the fœtus is expelled before the sixth month; after that it is premature birth. In law, however, any expulsion of the contents of the uterus before the full time is an abortion or miscarriage.

In deciding whether any substance expelled from the uterus is really a fœtus or a mole, and therefore the result of conception, or the coat of the uterus, and unconnected with pregnancy, the examination of the substances expelled must be carefully made. Moles are blighted fœtuses. An examination of the woman will be necessary, though it is not easy during the early months of pregnancy, and especially in those who have borne children, to say whether abortion has taken place or not. The history must be inquired into; the regular or exceptional use of drugs to promote menstruation is important, for in the former case no criminal intent may exist, although pregnancy be present. The state of the breasts, the hymen, and the os uteri, should all be carefully examined. Putting a few apparently unimportant questions as to the frequent use of purgatives, the presence or absence of constipation, will often assist the diagnosis as showing that the woman has acted in an unusual manner. Abortion may be procured by the introduction of instruments, by falls, violent exercise, blows on the abdomen, etc. In the hands of ignorant persons the use of instruments (sounds, bougies, skewers, etc.) is attended with great danger. Perforation of the vaginal walls, bladder, cervix, or uterus, may follow their use. Septic pelvic peritonitis may ensue, and the woman may lose her life. The person who has employed such means for inducing abortion is liable to be charged with the crime of murder. There is no evidence to show that ergot, savin, bitter-apple, pennyroyal, or any other drug administered internally, will cause a woman to abort, except when taken in such large doses that actual poisoning results, with inflammation of the contents of the true pelvis. In such cases reflex uterine contractions may be set up, and abortion may follow. Diachylon pills are largely employed to induce abortion, and very often the woman taking them suffers severely from lead-poisoning.

XXVI

INFANTICIDE

Infanticide, or the murder of a new-born child, is not treated as a specific crime, but is tried by the same rules as in cases of felonious homicide. The term is applied technically to those cases in which the mother kills her child at, or soon after, its birth. She is often in such a condition of mental anxiety as not to be responsible for her actions. It is usually committed with the object of concealing delivery, and to hide the fact that the girl has, in popular language, 'strayed from the paths of virtue.' The child must have had a separate existence. To constitute 'live birth,' the child must have been alive after its body was entirely born—that is, entirely outside the maternal passages—and it must have had an independent circulation, though this does not imply the severance of the umbilical cord. Every child is held in law to be born dead until it has been shown to have been born alive. Killing a child in the act of birth and before it is fully born is not infanticide, but if before birth injuries are inflicted which result in death after birth, it is murder. Medical evidence will be called to show that the child was born alive.

The methods of death usually employed are—(1) Suffocation by the hand or a cloth. (2) Strangulation with the hands, by a tape or ribbon, or by the umbilical cord itself. (3) Blows on the head, or dashing the child against the wall. (4) Drowning by putting it in the privy or in a bucket of water. (5) Omission: by neglecting to do what is absolutely necessary for the newly-born child—e.g., not separating the cord; allowing it to lie under the bed-clothes and be suffocated.

With regard to the question of the maturity of a child, the differences between a child of six or seven months and one at full term may be stated as follows:

Between the sixth and seventh month, length of child 10 to 14 inches—that is, the length of the child after the fifth month is about double the lunar months—weight 1 to 3 pounds; skin, dusky red, covered with downy hair (lanugo) and sebaceous matter; membrana pupillaris disappearing; nails not reaching to ends of fingers; meconium at upper part of large intestine; testes near kidneys; no appearance of convolutions in brain; points of ossification in four divisions of sternum.

At nine months, length of child 18 to 22 inches; weight, 7 to 8 pounds; skin rosy; lanugo only about shoulders; sebaceous matter on the body; hair on head about an inch long; testes past inguinal ring; clitoris covered by the labia; membrana pupillaris disappeared; nails reach to ends of fingers; meconium at termination of large intestine; points of ossification in centre of cartilage at lower end of femur, about 1-1/2 to 2-1/2 lines in diameter; umbilicus midway between the ensiform cartilage and pubis.

Owing to the difficulty of proving that the crime of infanticide has been committed, the woman may in England be tried for concealment of birth, and in Scotland for concealment of pregnancy, if she conceal her pregnancy during the whole time and fail to call for assistance in the birth. Either of these charges would only be brought against a woman who had obviously been pregnant, and now the child is missing or its dead body has been found. It is expected that every pregnant woman should make provision for the child about to be born, and so should have talked about it or have made clothes, etc., for it. The punishment for concealment is imprisonment for any term not exceeding two years. The charge of concealment is very often alternative to infanticide. To substantiate the charge, however, it must be proved that there had been a secret disposition of the dead body of the infant, as well as an endeavour to conceal its birth.

A woman may be delivered of a child unconsciously, for the contractile power of the womb is independent of volition. Under an anæsthetic the uterus acts as energetically as if the patient were in the full possession of her senses.

Nowadays a woman is rarely hanged for infanticide, and it is a mere travesty of justice to pass on her the death sentence, well knowing that it will never be executed.

XXVII

EVIDENCES OF LIVE BIRTH

The signs of live birth prior to respiration are negative and positive. A negative opinion may be formed when evidence is found of the child having undergone intra-uterine maceration. In this case the body will be flaccid and flattened; the ilia prominent; the head soft and yielding; the cuticle more or less detached, and raised into large bullæ; the skin of a red or brownish-red colour; the cavities filled with abundant bloody serum; the umbilical cord straight and flaccid.

A positive opinion is justified when such injuries are found on the body as could not have been inflicted during birth, and are attended with such hæmorrhage as could only have occurred while the blood was circulating. Fractures of the cranium from accidental falls (precipitate labour) are as a rule stellate, and are situated on the vertex or in the parietal protuberance. The fractures from violence are more extensive, usually depressed, and accompanied by laceration of the scalp.

The evidences of live birth after respiration has taken place are usually deduced from the condition of the lungs, though indications are also found in other organs. The diaphragm is more arched before than after respiration, and rises higher in the thorax in the former case than in the latter. The lungs before respiration are situated in the back of the thorax, and do not fill that cavity; they are of a dark, red-brown colour and of the consistence of liver, without mottling. After respiration they expand and occupy the whole thorax, and closely surround the heart and thymus gland. The portions containing air are of a light brick-red colour, and crepitate under the finger. The lungs are mottled from the presence of islands of aerated tissue, surrounded by arteries and veins. The weight of the lungs before respiration is about 550 grains, after an hour's respiration 900 grains; but this test is of little value. The ratio of the weight of the lungs to that of the body (Ploucquet's test), which is also unreliable, is, before respiration, about 1 to 70; after, 1 to 35. Lungs in which respiration has taken place float in water; those in which it has not, sink. There are exceptions to this rule, on which, however, is founded the hydrostatic test. As originally performed, this test consisted merely in placing the lungs, with or without the heart, in water, and noticing whether they sank or floated. The test is now modified by squeezing, and by cutting the lungs up into pieces.

The objections to the test as originally performed are—(1) That the lungs may sink as the result of disease—e.g., double pneumonia. (2) That respiration may have been so limited in extent that the lungs may sink, owing to large portions of lung tissue remaining unexpanded (atelectasis). (3) Putrefaction may cause the lungs to float when respiration has not taken place. (4) The lungs may have been inflated artificially. Few of these objections apply, however, when the hydrostatic test, modified by pressure, is employed. To take these objections in detail, it may be stated: (1) If the lungs sink from disease, the question of live birth is answered. (2) This objection is too refined for practical use. The lungs sink, there is an absence of any of the signs of suffocation, and the matter ends. The examiner has only to describe the conditions which he finds, and is not required to indulge in conjectures as to the amount of respiration which may or may not have taken place. (3) Gas due to putrefaction collects under the pleural membrane, and can be expelled by pressure, and is not found in the air cells. The lungs decompose late, hence in a fresh body putrefaction of the lungs is absent; in a putrefied child, if the lungs sink, it must have been stillborn. The so-called emphysema pulmonum neonatorum is simply incipient putrefaction.

The lung test simply shows that the child has breathed, but affords no proof that the child has been born alive. The child may have breathed as soon as its head protruded, the rest of the body being in the maternal passages. The child is not born alive until it has been completely expelled, although it is not necessary that the umbilical cord should have been cut.

In addition to these tests, live birth may be suspected from the following conditions: The stomach may contain milk or food, recognized by the microscope and by Trommer's test for sugar; the large intestines in stillborn children are filled with meconium, in those born alive they are usually empty; the bladder is generally emptied soon after birth; the skin is in a condition of exfoliation soon after birth. The organs of circulation undergo the following changes after birth, and the extent to which these changes have advanced will give an idea of how long the child has lived: The ductus arteriosus begins to contract within a few seconds of birth; at the end of a week it is about the size of a crow quill, and about the tenth day is obliterated. The umbilical arteries and vein: the arteries are remarkably diminished in calibre at the end of twenty-four hours, and obliterated almost up to the iliacs in three days; the umbilical vein and the ductus venosus are generally completely contracted by the fifth day. The foramen ovale becomes obliterated at extremely variable periods, and may continue open even in the adult.

Importance of late has been attached to the stomach-bowel test. If the stomach and duodenum contain air, and consequently float in water, the chances are that the child did not die immediately after birth; this is known as Breslau's second life test, and the lower the air in the intestinal canal, the greater is the probability that the child survived birth.

The umbilical cord in a new-born child is fresh, firm, round, and bluish in colour; blood is contained in its vessels. The cord may be ruptured by the child falling from the maternal parts in a precipitate labour, and the ruptured parts present ragged ends. It is seldom that a child bleeds to death from an untied or cut umbilical cord, and the chances in a torn cord are still more remote. The changes in the cord are as follows: First it shrinks from the ligature towards the navel; this change may begin early, and is rarely delayed beyond thirty hours; the cord becomes flabby, and there is a distinct inflammatory circle round its insertion. The next change is that of desiccation or mummification; the cord becomes reddish-brown, then flattened and shrivelled, then translucent and of the colour of parchment, and falls off about the fifth day. The third stage, that of cicatrization, then ensues about the tenth to the twelfth day. The bright red rim round the insertion of the cord, with inflammatory thickening and slight purulent secretion, may be considered as evidence of live birth, and the stage at which the separation of the cord by ulcerative process has arrived will point to the probable duration of time the child has existed after birth.

There are many fallacies in the application of any of these tests, and the whole subject bristles with difficulties. The medical witness would do well to exhibit a cautious reserve, for if the child dies immediately after birth it is almost impossible to prove that it was born alive.

XXVIII

CAUSE OF DEATH IN THE FŒTUS

The death of the fœtus may be due to—(1) Immaturity or intra-uterine malnutrition, or simply from deficient vitality; (2) complications occurring during or immediately after birth, which may either be unavoidable or inherent in the process of parturition, or may be induced with criminal intent.

In the latter category come such accidents as the pressure of tumours in the pelvic passages, or disease of the bones in the mother, or pressure on the cord from malposition of the child during labour, asphyxiation from the funis being twisted tightly round the neck or limbs, or from injuries due to falls on the floor in sudden labours. Where the death of the fœtus has been induced with criminal intent, it may be due to punctured wounds of the fontanelles, orbits, heart, or spinal marrow; dislocation of the neck;

separation of the head from the body; fracture of the bones of the head and face; strangulation; suffocation; drowning in the closet pan or privy, or from being thrown into water.

Under the head of infanticide by commission, we have injuries of all kinds; under infanticide by omission, neglecting to tie the cord, allowing it to be suffocated by discharges in the bed, neglect to provide food, clothes, and warmth, for the new-born child.

XXIX

DURATION OF PREGNANCY

The natural period of gestation is considered as forty weeks, ten lunar months, or 280 days. A medical witness would have to admit the possibility of gestation being prolonged to 300 days, and if this time were not very materially exceeded it would be well to give the woman the benefit of the doubt. It may be mentioned that 300 days is the extreme limit fixed by the French and Scottish law. No fixed period is assigned in English or American law to the duration of pregnancy, though it is allowed that utero-gestation may be greatly prolonged. In a recent case decided, the Lord Chancellor accepted a case where it was alleged pregnancy had extended to 331 days. A child only five months old may live, for a short time at all events. There is considerable difficulty in many cases in fixing the date of conception. The data from which it is calculated are the following: (1) Peculiar sensations attending conception, which are not sufficiently defined to be recognized by those conceiving for the first time. (2) Cessation of the catamenia. Other causes may, however, cause this; and, on the other hand, a woman may menstruate during the whole period of her pregnancy. This datum also gives a variable period, and may involve an error of several days or a month, for the menses may be arrested by cold, etc., at one monthly period, and the woman become pregnant before the next. (3) The period of quickening. This, when perceived (which is not always the case), also occurs at variable periods from the tenth to the twenty-sixth week. (4) A single coitus. This does not, however, correspond to the time of fertilization. Several days may elapse before the spermatozoa meet with an ovum and fertilize it.

In Scotland a child born six months after marriage is legitimate, which is allowing an ample margin.

XXX

VIABILITY OF CHILDREN

A child may be born alive, but may not be viable, by which is meant that it is not endowed with a capacity of maintaining its life. Speaking generally, 180 days represents the lowest limit at which a child is viable, but prolonged survival under these circumstances is the exception. Many cases, however, have been recorded in which children born at six months have been reared. The signs of immaturity and maturity may be thus tabulated:

Immaturity.	Maturity.
Centre of body high; head disproportionate in size; membrana pupillaris present; testicles undescended; deep red colour of parts of generation; intense red colour, mottled appearance, and downy covering, of skin; nails not formed; feeble movements;	Strong movements and cries as soon as born; body clear, red colour, coated with sebaceous matter; mouth, nostrils, eyelids, and ears, open; skull somewhat firm, and fontanelles not far apart; hair, eyebrows, and nails, perfectly developed; testicles descended; free discharge of urine

inability to suck; necessity for artificial heat; almost unbroken sleep; rare and imperfect discharges of urine and meconium; closed state of mouth, eyelids, and nostrils. and meconium; power of suction, indicated by seizure on the nipple or a finger placed in the mouth.

XXXI

LEGITIMACY

A child born in wedlock is presumed to have the mother's husband for its father. This may, however, be open to question upon the following grounds: Absence or death of the reputed father; impotence or disease in the husband preventing matrimonial intercourse; premature delivery in a newly-married woman; want of access; and the marriage of the woman again immediately on the death of her husband. In the last case, where either husband might have been the father, the child at the age of twenty-one is at liberty to select its father from the possible pair.

A child born of parents before marriage is in Scotland rendered legitimate by their subsequent marriage, but in England the offspring remains illegitimate whether the parents marry or not after its birth. The offspring of voidable or invalid marriages may be made legitimate by application to the courts.

There is a difference between being legitimate and lawfully begotten. A child born in wedlock is legitimate, but if the parents were married only a week previously it could not have been lawfully begotten.

The Acts and rulings relating to Marriage and Legitimacy are extremely complicated. It is not putting it too strongly to say that a very large number of people in this country who believe themselves to be legally married are not married at all, and that thousands of children who have not the slightest doubt as to their legitimacy are in the eyes of the law bastards.

XXXII

SUPERFŒTATION

By superfœtation is meant the conception, by a woman already pregnant, of a second embryo, resulting in the birth of two children at the same time, differing much in their degree of maturity, or in two separate births, with a considerable interval between. The possibility of the occurrence of superfœtation has been doubted, but there are well-authenticated cases which countenance the theory of a double conception. It has been shown that the os uteri is not closed, as was once supposed, immediately on conception. Should an ovum escape into the uterus, it may become impregnated a month or so after a previous conception. The most probable explanation is that the case has been one of twins, one being born prematurely; or, on the other hand, the uterus may have been double, and conception may have taken place in one cornu at a later period than in the other cornu.

XXXIII

INHERITANCE

In order to inherit, the child must be born alive, must be born during the lifetime of the mother, and must be born capable of inheriting—that is to say, monsters are incapable of inheriting. There is a mode of inheritance called 'tenancy by courtesy.' When a man marries a woman possessed of an estate or inheritance, and has, by her, issue born alive in her lifetime capable of inheriting her estate, in this case he shall, on the death of his wife, hold the lands for his life as tenant by the courtesy of England. The meaning of the words 'born alive' in this instance is not the same as in cases of infanticide. In Civil law any motion of the child's body, however slight, or the fact of it having been heard to cry by witnesses, is held to be sufficient proof of the child having been born alive. It may die immediately afterwards, and it is not necessary that the child be viable.

XXXIV

IMPOTENCE AND STERILITY

In the male, impotence may arise from physical or mental causes. The physical causes may be—too great or too tender an age; malformation of the genital organs; crypsorchides, defect or disease in the testicles; constitutional disease (diabetes, neurasthenia, etc.); or debility from acute disease, as mumps. Masturbation, and early and excessive sexual indulgence, are also causes. The mental causes include—passion, timidity, apprehension, aversion, and disgust. The case will be remembered of the man who was impotent unless the lady were attired in a black silk dress and high-heeled French kid boots.

If a man is impotent when he marries, the marriage may be set aside on the ground that it had never been consummated. The law requires that the impotency should have existed ab initio—that is, before marriage—and should be of a permanent or incurable nature; marriage, as far as the law goes, being regarded as a contract in which it is presupposed that both the contracting parties are capable of fulfilling all the objects of marriage. In the case of the Earl of Essex the defendant admitted the charge as regards the Countess, but pleaded that he was not impotent with others, as many of her waiting-maids could testify. When a man becomes impotent after marriage, his wife must accept the situation, and has no redress. A man may be sterile without being impotent, but the law will not take cognizance of that. The wife may be practically impotent, but the law will not assist the husband. He must continue to do his best under difficult circumstances. In former times in case of doubt a husband was permitted to demonstrate his competency in open court, but this custom is no longer regarded with favour by the judges.

The removal of the testicles does not of necessity render a man impotent, although it deprives him of his procreative power. Eunuchs are capable of affording illicit pleasure, whilst the male sopranos, or castrati, are often utilized for that purpose.

In the female, impotence may be caused by the narrowness of the vagina, adhesion of the vulva, absence of vagina, imperforate hymen, and tumours of the vagina.

Sterility in women may occur from the above-named causes of impotence, together with absence of the uterus and ovaries, or from great debility, syphilis, constant amenorrhœa, dysmenorrhœa, or menorrhagia.

XXXV

RAPE

Rape is the carnal knowledge of a woman by force and against her will. The resistance of the woman must be to the utmost of her power, but if she yield through fear or duress it is still rape. The woman is a competent witness, but her statements may be impugned on the ground of her previous bad character, and evidence may be called to substantiate the charge. The perpetrator must be above the age of fourteen years.

The definition of rape which we have given is not altogether satisfactory. Take, for example, the case of a woman who goes to bed expecting her husband to return at a certain hour. The lodger, let us say, takes advantage of this fact, and, getting into bed, has connection with her, she not resisting, assuming all the while that it is her husband. This is rape, but it is not 'by force,' and it is not 'against her will,' but it is 'without her consent,' as she has not been fully informed as to all the circumstances of the case.

In all cases of rape in which there is no actual resistance or objection, consent may be assumed. It is not essential that the woman should state in so many words that she does not object. The force used may be moral and not physical—e.g., threats, fear, horror, syncope.

By 48 and 49 Vict., c. 49, the carnal knowledge of a girl under thirteen is technically rape. The consent of the girl makes no difference, since she is not of an age to become a consenting party.

An attempt at carnal knowledge of a girl under thirteen is a misdemeanour. Her consent makes no difference, and even the solicitation of the act on the part of the child will not exonerate the accused.

Intercourse with a girl between thirteen and sixteen, even with her consent, is a misdemeanour.

This Act is a favourite with the blackmailer. The child is sent out to solicit, dressed like a woman, but appears in the witness-box in a much more juvenile costume.

To constitute rape there must be penetration, but this may be of the slightest. There may be a sufficient degree of penetration to constitute rape without rupturing the hymen. Proof of actual emission is now unnecessary.

The subject of carnal knowledge (C.K.) or its attempt may be summed up as follows:

Under thirteen	C.K.	Felony.
Under thirteen	Attempt	Misdemeanour.

Consent no defence.

From thirteen to sixteen	C.K.	Misdemeanour.
From thirteen to sixteen	Attempt	Misdemeanour.

Consent and even solicitation no defence.

Reasonable cause to believe the girl over sixteen is a good defence.

Charge must be brought within three months.

Over sixteen	C.K. with consent	Nil.

Subject to civil action for loss of girl's services by father.

Idiot or imbecile	C.K. with violence	Rape.
Idiot or imbecile	C.K. without violence	Misdemeanour.
Personation of husband		Rape.

Tacit consent no defence, for obtained by fraud.

Married woman	C.K. with consent	Adultery.
Mother, sister, daughter, granddaughter	C.K. consent immaterial; born in wedlock or not	Incest.
Females	Indecent assaults	Misdemeanour.

It is a misdemeanour to give to a woman any drug so as to stupefy her, and so enable any person to have unlawful connection with her.

False charges of rape are very often made. The motive may be to extort blackmail, revenge, or mere delusion. On examining such cases bruises are seldom found, but scratches which the woman has made on the front of her body may be discovered, and the local injuries to the generative organs are slight, if present at all.

Physical Signs.—In the adult the hymen may be ruptured, the fourchette lacerated, and blood found on the parts, together with scratches and other marks and signs of a struggle. In the child there may be no hæmorrhage, but there will be indications of bruising on the external organs, with probably considerable laceration of the hymen, the laceration in some cases extending into the rectum. Severe hæmorrhage, and even death, may follow the rape of a young child. The patient will have difficulty in walking, and in passing water and fæces. After some hours the parts are very tender and swollen, and a sticky greenish-yellow discharge is present. These signs last longer in children than in adults; but as a rule—in the adult, at least—all signs of rape disappear in three or four days. Young and delicate children may suffer from a vaginal discharge, with swelling of the external genitals, simulating an attempt at rape. Infantile leucorrhœa is common, and many innocent people have been exposed to danger from false charges of rape on children, instituted as a means of levying blackmail. A knowledge of these facts suggests the necessity of giving a guarded opinion when children are brought for examination in suspected cases. Pregnancy may follow rape.

Seminal stains render the clothing stiff and greyish-yellow in colour, with translucent edges. On being moistened they give the characteristic seminal odour.

Semen may be found on the linen of the woman and man, and will be recognized under the microscope by the presence in it of spermatozoa, minute filamentary bodies with a pear-shaped head; but it must not be forgotten that the non-detection of spermatozoa is no proof of absence of sexual intercourse, for these bodies are not always present in the semen of even healthy adult young men. Spermatozoa must not be mistaken for the Trichomonas vaginæ found in the vaginæ of some women. The latter have cilia surrounding the head, which is globular.

Florence's Micro-Chemical Test for Spermatic Fluid.—If a drop of the fluid obtained by wetting a supposed spermatic stain be mixed with a drop of the following solution (KI, parts 1.65; pure iodine, 2.54; distilled water, 30) in a watch-glass, brownish-red pointed crystals resembling hæmin crystals are obtained.

Barberio's Test.—Mix a drop of the spermatic stain with a drop of a saturated solution of picric acid, when needle-shaped yellow rhombic crystals are formed.

Gonorrhœal Stains.—A cover-glass preparation stained with methylene blue reveals the gonococci lying in pairs within the leucocytes.

XXXVI

UNNATURAL OFFENCES

Trials for sodomy and bestiality are common at the assizes, but, as they are rarely reported, they fail to attract attention. Sodomy is a crime both in the active and passive agent, unless the latter is a non-consenting party. The evidence of either associated may be received as against his colleague. If the crime is committed on a boy under fourteen, it is a felony in the active agent only. As in cases of rape, emission is not essential, and penetration, however slight, answers all practical purposes.

There can be no doubt that in the majority of these cases there exists a congenitally abnormal condition of the sexual instinct, these individuals from their childhood manifesting a perverted sexual instinct. The man is physically a man, but psychically a woman, and vice versâ. The tendency nowadays is not to charge these people with the more serious offence, but to deal with them under Section 11 of the Criminal Law Amendment Act, 1885 (48 and 49 Vict., c. 69). This section, which is sufficiently comprehensive, runs as follows: 'Any male person who in public or private commits or is a party to the commission, or attempts to procure the commission by any male person, of any act of gross indecency with another male person, shall be guilty of a misdemeanour.' The penalty is imprisonment for two years, with or without hard labour. It is provided by Section 4 of the same Act that a boy under sixteen may be whipped.

Incest.—This crime is dealt with under the Punishment of Incest Act, 1908 (8 Edward VII., c. 45). Carnal knowledge with mother, sister, daughter, or grand-daughter, is a misdemeanour, provided the relationship is known. It also applies to the half-brother and half-sister. It is equally an offence whether the relationship can or cannot be traced through lawful wedlock. Consent is no defence. A woman may be charged under the Act if she, being above the age of sixteen, with consent permits her grandfather, father, brother, or son, to have carnal knowledge of her.

XXXVII

BLACKMAILING

There are in London and every large city scores of men and women who live by blackmailing or chantage. There are many different forms of this industry. There is the man who knows something about your past life, which he threatens to reveal to your friends or colleagues unless you buy him off. There is the breach-of-promise blackmailer, and there is the female patient, who threatens to charge you with improper conduct or indecent assault. Medical men from their position are often selected as victims. The introduction of corridor carriages on many of our railways has done much to stamp out one particular form of blackmailing, but public urinals are still a source of danger.

It is the worst possible policy to temporize with a blackmailer. If you give him a single penny, you are his for life. It is as well to remember that it is just as criminal to attempt to extract money from a guilty as from an innocent person. It is of no use attempting to deal with these cases single-handed. You must not only deny the allegation, but 'spurn the allegator.' Put the matter into the hands of a good sharp criminal solicitor, and instruct him to rid you of the nuisance by taking criminal proceedings.

XXXVIII

MARRIAGE AND DIVORCE

Marriage may be accomplished in many ways: (1) By the publication of banns; (2) by an ordinary licence; (3) by a special licence; (4) by the Superintendent-Registrar's licence; (5) by a special licence granted by the Archbishop of Canterbury in consideration of the payment of the sum of £25. Then, for persons having a domicile in Scotland, there is the marriage by repute. The consent of the parties, which is the essence of the contract, may be expressed before witnesses, and it is not requisite that a clergyman should assist, but it is essential that the expressions of consent must be for a matrimonial intent. 'Habit and repute' constitute good evidence, but the repute must be the general, constant, and unvarying belief of friends and neighbours. The cohabitation must be in Scotland.

Any irregularity in the marriage ceremony or the non-observance of any formality will not invalidate the marriage, unless it were known to both the contracting parties. If a man were married in a wrong name the contract would still be valid if the wife were unacquainted with the deception at the time. If the person who officiated were a bogus clergyman, the marriage would hold good if the contracting parties supposed him to be a properly ordained priest. In a case in which a marriage was solemnized in a building near the church at a time when the church was undergoing repairs, and where during such alterations Divine service had been performed, it was held that the ceremony was good. To all intents and purposes marriage comes under the 'Law of Contract' (see Anson, W.R., Bart.), and the law looks to the intention rather than to the actual details. All marriages between persons within the prohibited degrees of consanguinity or affinity are null and void. This prohibition extends both to the illegitimate as well as the legitimate children of the late wife's or husband's parents. A marriage with a deceased wife's sister is now legal in Great Britain and the Colonies, and is recognized in most foreign countries. A common device with people within the prohibited degrees is to get married abroad, but such marriage is strictly speaking inoperative, and the children of such union are illegitimate. Practically, however, it is a matter of no importance, for when people live together and say they are married, they are accepted at their own estimate.

A man can obtain a divorce from his wife if he can prove that she has been guilty of adultery since her marriage. This may be established by inference. Obviously, it is difficult in the majority of cases to establish by ocular demonstration that adultery has been committed. But given evidence of familiarity and affection with opportunity and suspicious conduct, a jury will commonly infer it.

A woman cannot obtain a divorce from her husband for adultery alone. She must prove adultery plus cruelty, or adultery plus desertion without reasonable cause. Failing this, she may be able to prove either bigamy or incestuous adultery. Legal cruelty is a very comprehensive term, and does not of necessity mean physical violence. If the husband as the result of his infidelity were to give his wife a contagious disease, that would constitute cruelty. Taking a more extreme case, if a husband were to have connection in her house with his wife's maid, that would probably be held to constitute cruelty, as it would tend to lower her in the eyes of her servants.

A wife can obtain a judicial separation if she can prove (1) adultery, (2) cruelty, or (3) desertion without reasonable cause for two years. If a husband is away on his business, as, for example, the case of an officer ordered abroad, that is not desertion. For a woman to get a judicial separation, it is sufficient if she can prove one variety of matrimonial offence, but for a divorce she requires more than one.

The jury may find that Mrs. A. has committed adultery with Mr. B., but that Mr. B. has not committed adultery with Mrs. A. The explanation is, that a wife's confession is evidence against herself, but not against another person. You can confess your own sins, but not another's.

The Divorce Law of Scotland differs materially from that of England. In Scotland there is no decree nisi, no decree absolute, and no intervention by the King's Proctor. Instead there is a single and final judgment, and when a decree of divorce is pronounced the successful litigant at once succeeds to all rights, legal and conventional, that would have come to him or her on the death of the losing party. If the husband is the offender, the wife in such circumstances may claim her right to one-third of his real estate; and if there are children, to one-third of his personal property, and to one-half if there are none.

Voidable Marriages.—If a man and woman go through the marriage ceremony, such a contract is null and void under the following circumstances: (1) Where bigamy has been committed; (2) if one of the parties were insane at the time of marriage; (3) where the plaintiff is under sixteen years of age; (4) when the marriage has not been consummated or followed by cohabitation; (5) when one of the parties was incapable of performing the marital act (impotent, and such not known by the other at the time); (6) when drunkenness had been induced so as to obtain consent; (7) concealment of pregnancy at the time of marriage.

XXXIX

FEIGNED DISEASES

Malingering in its various forms is by no means uncommon, and by many is regarded as a disease in itself. It is necessary, however, to distinguish between those cases in which it is feigned for some definite purpose—for example, to escape punishment or avoid public service—and those in which there is adequate motive, and the patient shams simply with the view of exciting sympathy, or from the mere delight of giving trouble. It is not uncommon for individuals summoned on a jury, or to give evidence in the law courts, to apply to their doctor for a certificate, assigning as a cause of exemption neuralgia, or some similar complaint unattended with objective symptoms. In such cases it is well to remind the patient that in most courts such certificates are received with suspicion, and are often rejected, and that the personal attendance of the medical man is required to endorse his certificate on oath.

Malingering has become much more common since the National Health Insurance Act has been passed. The possibility of obtaining a fair sum each week without the necessity of working for it induces many persons either to feign disease or to make recovery from actual disease or accident much more tedious than it ought really to be.

The feasibility of successfully malingering is greatly enhanced by the possession of some chronic organic disease. An old mitral regurgitant murmur is useful for this purpose.

It is not flattering to one's vanity to overlook a case of malingering, but should this occur little harm is done. It is a much more serious matter to accuse a person of malingering when in reality he may be suffering from an organic disease.

Here are some of the diseases which are most frequently feigned:

Nervous Diseases, as headache, vertigo, paralysis of limbs, vomiting, sciatica, or incontinence or suppression of urine, spitting of blood; others, again, simulate hysteria, epilepsy, or insanity.

On the other hand, the malingerer may actually produce injuries on his person either to excite commiseration or to escape from work. Thus, the beggar produces ulcers on his legs by binding a penny-piece tightly on for some days; the hospital patient, in order to escape discharge, produces factitious skin diseases by the application of irritants or caustics.

It is much more difficult to decide whether certain symptoms are due to a real disease which is present, or whether they are merely exaggerations of slight symptoms or simulations of past ones. The miner, after an injury to his back, recovers very slowly, if at all. He is suffering from 'traumatic neurasthenia'—a condition only too often simulated, and a disease very difficult to diagnose accurately. The miner takes advantage of our ignorance, and continues to draw his compensation. A workman during his work receives a fracture; instead of being able to resume work in six weeks, he asserts that the pain and stiffness prevent him, and this disability may persist for months. Such cases as these frequently come before the courts when the employer has discontinued to pay the weekly compensation for the injury. Medical men are called to give evidence for or against the injured workman.

Epilepsy is often simulated. The foaming at the mouth is produced by a piece of soap between the gums and the cheek. The true epileptic, especially if he suspects that a fit is imminent, takes his walks abroad in some secluded spot, whilst the impostor selects a crowded locality for his exertions. The epileptic often injures himself in falling, his imitator never; one bites his tongue, but the other carefully refrains from doing so. The skin of an epileptic during an attack is cold and pallid, but that of the exhibitor is covered with sweat as the result of his exertions. In epilepsy the urine and fæces are passed involuntarily, but his colleague rarely considers it necessary to carry his deception to this extent. In true epilepsy the eyes are partly open, with the eyeballs rolling and distorted, whilst the pupils are dilated and do not contract to light; the impostor keeps his eyes closed, and he cannot prevent the iris from contracting when a bicycle-lamp is flashed across his face. A useful test is to give the impostor a pinch of snuff, which promptly brings the entertainment to an end.

Lumbago is often feigned, and the imposture should be suspected when there is a motive, and when physical signs, such as nodes and tender spots, are absent. A simple test is to inadvertently drop a shilling in front of him, when he will promptly stoop and pick it up. The same principles apply to spurious sciatica.

Hæmorrhages purporting to come from the lungs, stomach, or bowels, rarely present much difficulty. The microscope is of use in all cases of bleeding. Possibly the gums or the inside of the cheeks may have been scratched or abraded with a pin.

Skin Diseases are excited artificially, especially those which may be produced by mechanical and chemical irritants. The most commonly employed are vinegar, acetic acid, carbolic acid, nitric acid, and carbonate of sodium; but tramps frequently use sorrel and various species of ranunculus. The lesions simulated are usually inflammatory in character, such as erythema, vesicular and bullous eruptions, and ulceration of the skin. They may be complicated by the presence of pediculi and other animal and vegetable parasites. Chromidrosis of the lower eyelids in young women often owes its origin to a box of paints. Factitious skin diseases are seen most commonly on the face and extremities, especially on the left side— in other words, on the most accessible parts of the body.

Feigned menstruation, pregnancy, abortion, and recent delivery are common, and should give rise to no difficulty. The same may be said of feigned insanity, aphonia, deaf-mutism, and loss of memory.

The following hints may be useful to a medical man when called to a supposed case of malingering: Do not be satisfied with one visit, but go again and unexpectedly; see that the patient is watched between the visits; make an objective examination, compare the indications with the statements of the patient, noting especially any discrepancies between his account of his symptoms and the real symptoms of disease; ask questions the reverse of the patient's statements, or take them for granted, and he will often be found to contradict himself; have all dressings and bandages removed; suggest, in the hearing of the patient, some heroic methods of treatment—the actual cautery, or severe surgical operation, for example; finally, chloroform will be found of great use in the detection of many sham diseases.

XL

MENTAL UNSOUNDNESS

The presumption in law is in favour of a person's sanity, even though he may be deaf, dumb, or blind.

The terms 'insanity,' 'lunacy,' 'unsoundness of mind,' 'mental derangement,' 'madness,' and 'mental alienation or aberration,' are indifferently applied to those states of disordered mind in which the person loses the power of regulating his actions and conduct according to the ordinary rules of society. The reasoning power is lost or perverted, and he is no longer fitted to discharge those duties which his social position demands. In some cases of insanity, as in confirmed idiocy, there is no evidence of the exercise of the intellectual faculties. It is probable that no standard of sanity as fixed by nature can be said to exist. The medical witness should decline to commit himself to any definition of insanity. There is no practical advantage in attempting to classify the different forms of insanity.

According to English law, madness absolves from all guilt, but in order to excuse from punishment on this ground it must be proved that the individual was not capable of distinguishing right from wrong in relation to the particular act of which he is accused, and that he did not know at the time of committing the crime that the offence was against the laws of God and nature.

Lunatics are competent witnesses in relation to testimony, as in relation to crime, if they understand the nature of an oath and the character of the proceedings in which they are engaged. The judge, as in the case of children, examines the lunatic tendered as a witness as to his knowledge of the nature and obligation of an oath, and, if satisfied, he allows him to be sworn.

A person, if suffering from such a state of mental unsoundness as to be unable to take care of his property, may be placed under the care of the Court of Chancery. The Court then administers his property, and otherwise allows him entire freedom of action.

With regard to the care of lunatics, no person is allowed to receive more than one lunatic into his house unless such house is licensed and the proper certificates have been signed. One patient may be taken without the house being licensed, but the usual certificates must in all cases be signed, and the Lunacy Commissioners communicated with. If a person receives another not of unsound mind into his house, and such person becomes subsequently insane, the person so keeping him renders himself liable to heavy penalties, unless the legal certificates are at once procured and the Commissioners of Lunacy communicated with.

At common law it appears that a lunatic cannot be placed in an asylum unless dangerous to himself or to others, but under the Lunacy Acts the placing of a madman in an asylum is considered as a part of the treatment with a view to the cure of the patient.

XLI

IDIOCY, IMBECILITY, CRETINISM

Idiocy is not a disease, but a congenital condition in which the intellectual faculties are either never manifested or have not been sufficiently developed to enable the idiot to acquire an amount of knowledge equal to that acquired by other persons of his own age and in similar circumstances with himself. Idiots, as a rule, are deformed in body as well as deficient in mind. Their heads are generally small and badly-shaped, and their features ill-formed and distorted. The teeth are few in number and very irregular. The hard palate has a very deep arch, or may even be cleft. The complexion is sallow and unhealthy, the limbs imperfectly developed, and the gait is awkward, shambling, and unsteady. In his legal relations an absolute idiot is civilly disabled and irresponsible, but in regard to crime, or as a witness, see remarks made above.

Imbecility is a form of mental defect not usually congenital, but commencing in infancy or in early life. The line of demarcation between the imbecile and the idiot may be found in the possession by the former of the faculty of speech, in distinction from the mere parrot-like utterance of a few words which can be taught the idiot. Imbecility may be intellectual, moral, or general. Questions frequently arise as to their responsibility for actions done by them, or as to their ability to manage their own affairs.

Cretinism is a form of amentia, which is endemic in certain districts, especially in some of the valleys of Switzerland, Savoy, and France. The malady is not congenital, but its symptoms usually appear within a few months of birth. The characteristics of this form of idiocy are an enlarged thyroid gland constituting a goitre or bronchocele, a high-arched palate, dwarfed stature, squinting eyes, sallow complexion, small legs, conical head, large mouth, and indistinct speech.

Feeble-Minded.—These are persons who are capable of earning a living under favourable circumstances, but are incapable, from mental defect which has existed from birth or from an early age, of (a) competing on equal terms with their normal fellows, or (b) of managing themselves and their affairs with ordinary prudence. Feeble-mindedness may affect the moral nature only, rendering the person selfish, untruthful, obscene, or unemployable. The Act of 1899 controls feeble-minded children; many such become paupers, criminals, prostitutes, etc.

Mental Deficiency and Lunacy Act, 1913.—Those included under this Act are idiots, imbeciles, feeble-minded persons, and moral imbeciles. The parents or guardians of such children between the ages of five and sixteen years must provide for them education and proper care. If they are unable to do so, the School Boards or Parish Councils must do so.

XLII

DEMENTIA: ACUTE, CHRONIC, SENILE, AND PARALYTIC

In dementia the mental aberration does not occur until the mind has become fully developed, thus differing from amentia, which is congenital or comes on very early in life.

Acute Dementia.—This is a condition of profound melancholy or stupor, which arises from sudden mental shock, the mind being, as it were, arrested and fixed in abstraction on the event.

Chronic Dementia is generally caused by the gradual action on the mind of grief or anxiety, by severe pain, mania, apoplexy, paralysis, or repeated attacks of epilepsy.

Senile Dementia is a form which is incidental to aged persons, and commences gradually with such symptoms as loss of memory for recent events, dulness of perception, and inability to fix the attention. Later on the reasoning powers begin to fail, and finally, memory, reason, and power of attention, are quite lost, the muscular power and force remaining intact. In the last stage there is simply bare physical existence.

General Paralysis of the Insane, Paralytic Dementia.—This is a most interesting form of dementia. It is closely allied to, if not identical with, locomotor ataxy. Its most prominent and characteristic symptom consists in delusions of great power, exalted position, and unlimited wealth—megalomania. The exaltation is universal, and the patient may maintain at one and the same time that he is running a theatrical company, that he is the Prince of Wales, and that he is the Almighty. Moral perversion is a common symptom, and the patient is often guilty of criminal assaults, indecent exposures, bigamous marriages, and the like. It is accompanied with progressive bodily and mental decay. Women are comparatively rarely affected by it, and it generally commences in men about middle age, and its duration is from a few months to three years. It is commonly parasyphilitic in origin. Paralytic symptoms first appear in the tongue, lips, and face; the speech becomes thick and hesitating. The paralytic symptoms gradually go on increasing, the sphincters refuse to act, and death may occur from suffocation and choking. Sometimes, during the earlier stages especially, there may be maniacal paroxysms or epileptic fits. The delusions remain the same throughout, the patient always expresses himself as being happy, and his last words will probably have reference to money and other absurd delusions.

When a person of hitherto blameless life is charged with an act of indecency, he should be examined for G.P.I. The condition of his prostate should also be investigated. He may be suffering from either mental or physical disease, or both (see p. 59).

XLIII

MANIA

Under the term 'mania' are included all those forms of mental unsoundness in which there is undue excitement. It is divided into general, intellectual, and moral, and each of the two latter classes again into general and partial.

General Mania affects the intellect as well as the passions and emotions. Mania is usually preceded by an incubative period in which the patient's general health is affected. The duration of this period may vary from a few days to fifteen or twenty years. When the disease is established, the patient has paroxysms of violence directed against himself as well as others. He tears his clothes to pieces, either abstains from food and drink or eats voraciously, and sustains immense muscular exertion without apparent fatigue. The face becomes flushed, the eye wild and sparkling; there is pain, weight, and giddiness in the head, with restlessness.

General Intellectual Mania, attacking the intellect alone, is rare; but some one emotion or passion, as pride, vanity, or love of gain, may obtain ascendancy, and fill the mind with intellectual delusions.

A delusion may be defined as a perversion of the judgment, a chimerical thought; an illusion, an incorrect impression of the senses, counterfeit appearances; hence we speak of a delusion of the mind, an illusion of the senses. Lawyers lay great stress on the presence of delusions as indicative of insanity. An hallucination is a sensation which is supposed by the patient to be produced by external impressions, although no material object acts upon his senses at the time.

Partial Intellectual Mania, or **Monomania**, also called **Melancholia**, is a form of the disease in which the patient becomes possessed of some single notion, contradictory alike to common-sense and his own experience.

General Moral Mania.—This is a morbid perversion of the natural feelings, affections, inclinations, temper, habits, moral dispositions, and natural impulses, without any remarkable disorder or defect of the intellect, or knowing and reasoning faculties, and particularly without any insane illusion or hallucination. It is often difficult to distinguish this form of mania from the moral depravity which we associate with the criminal classes.

Partial Moral Mania—Paranoia—Delusional Insanity.—In this form one or two only of the moral powers are perverted. Delusions are always present, and very frequently are those of persecution. The patient's conduct is dominated by his delusion; thus murder and suicide may be committed. There are several forms:

Kleptomania, a propensity to theft; common in women in easy circumstances. Dipsomania, or Oinomania, an insatiable desire for drink. Morphinomania, a craving for morphine or its preparations. Erotomania, or amorous madness. When occurring in women this is also called Nymphomania, and in men Satyriasis. It consists in an uncontrollable desire for sexual intercourse. Pyromania, an insane impulse to set fire to everything. Homicidal mania, a propensity to murder. Suicidal mania, a propensity to self-destruction. Some consider suicide as always a manifestation of insanity.

Insanity of Pregnancy.—This may show itself after the third month of pregnancy in the form of melancholia. It is not recovered from until after delivery.

Puerperal Mania.—This form of mania attacks women soon after childbirth. There is in many cases a strong homicidal tendency against the child.

Insanity of Lactation comes on four to eight months after parturition, either as mania or melancholia. The mother may repeatedly attempt suicide.

Mania with Lucid Intervals.—In many cases mania is intermittent or recurrent in its nature, the patient in the interval being in his right mind. The question of the presence or absence of a lucid interval frequently occurs where attempts are made to set aside wills made by persons having property. In these cases the law, from the reasonableness of the provisions of the will, may assume the existence of the lucid interval. A will made during a lucid interval is valid. When an attempt is made to set aside the provisions of a will on the ground of insanity in a person not previously judged insane, the plaintiff must show that the testator was mad; when the provisions of the will of a lunatic are attempted to be upheld, the plaintiff must show that the will was made during a lucid interval.

A testator is capable of making a valid will when he has (1) a knowledge of his property and of his kindred; (2) memory sufficient to recognize his proper relations to those about him; (3) freedom from delusions affecting his property and his friends; and (4) sufficient physical and mental power to resist undue influence.

The fact of a man being subject to delusions may not affect his testamentary capacity. He may believe himself to be a tea-kettle, and yet be sufficiently sound mentally to make a valid will.

Undue Influence.—Persons of weak mind or those suffering from senile dementia are often said to have been unduly influenced in making their wills, and subsequently their dispositions are disputed in court. Before witnessing the will made by such a person, the medical man should satisfy himself that the testator is of a 'sound disposing mind.' This he will do by questioning, and his knowledge of the home-life of the patient will either confirm or set aside the idea of influence.

A person who is aphasic may be competent to make a will. He may not be able to speak, but may understand what is said to him, and may be able to indicate his wishes by nods and shakes of the head. Ask him if he wishes to make a will, then inquire if he has £10,000 to leave, then if he has £100, and in this way arrive approximately at the sum. Then ask him if he wishes to leave it all to one person. If he nods assent, ask if it be to his wife or some other likely person. If he wishes to divide it, ascertain his intention by definite questions, and, having ascertained his views, commit them to writing, read the document over to him, and ask if it expresses his intentions. That being settled, a mark which he acknowledges in the presence of two witnesses, preferably men of standing, will constitute a valid document.

In certain forms of neurasthenia, the 'phobias' are common, but must not be regarded as evidence of insanity. 'Agoraphobia' is the fear of crossing an open space, 'batophobia' is the fear that high things will fall, 'siderophobia' is the fear of thunder and lightning, 'pathophobia' is the fear of disease, whilst 'pantophobia' is the fear of everything and everybody.

Epilepsy in Relation to Insanity.—The subjects of this disease are often subject to sudden fits of uncontrollable passion; their conduct is sometimes brutal, ferocious, and often very immoral. As the fits increase in number, the intellect deteriorates and chronic dementia or delusional insanity may supervene. (1) Before a fit the patient may develop paroxysms of rage with brutal impulses (preparoxysmal insanity), and may commit crimes such as rape or murder. (2) Instead of the usual epileptic fit, the patient may have a violent maniacal attack (masked epilepsy, epileptic equivalent, psychic form of epilepsy). (3) After the fit the patient may perform various automatic actions (post-epileptic automatism) of which he has no subsequent recollection. Thus the patient may urinate or undress in a public place, and may be arrested for indecent exposure. Epileptics who suffer from both petit and grand mal attacks are specially liable to maniacal attacks. Such insanity differs from ordinary insanity in its sudden onset, intensity of symptoms, short duration and abrupt ending. To establish a plea of epilepsy in cases of crime, one must show that the individual really did suffer from true epilepsy, and that the crime was committed at a period having a definite relation to the epileptic seizure.

Alcoholic Insanity.—This may occur in three forms:

1. *Acute Alcoholic Delirium (mania a potu)*, due to excessive amount of alcohol consumed.

2. *Delirium Tremens*, due to long-continued over-drinking. The patient suffers from horrible dreams, illusions, and suspicions, which may lead him to attack people or commit suicide.

3. *Chronic Alcoholic Insanity*. Loss of memory is the chief symptom, with paralysis of motion, hallucinations and delusions of persecution.

Responsibility for Criminal Acts.—To establish a defence on the ground of insanity, it must be proved that the prisoner at the time when the crime was committed did not know the nature and quality of the act he was committing, and did not know that it was wrong. At the present time, however, the power of controlling his actions is usually made the test.

The plea of insanity is brought forward, as a rule, only in capital charges, so that the prisoner, if found guilty, will escape hanging. If proved 'guilty, but insane,' the person is sentenced to be kept in a criminal lunatic asylum 'during His Majesty's pleasure.'

XLIV

EXAMINATION OF PERSONS OF UNSOUND MIND

The following hints with regard to the examination of patients supposed to be insane will be useful: The general appearance and shape of head, complexion, and expression of countenance, gait, movements, and speech, should be noted; the state of the general health, appetite, bowels, tongue, skin, and pulse, should be inquired into; and in women the state of the menstrual function should be ascertained. The family history must be traced out, and the personal history taken with care, especially as to whether the unsoundness came on late in life or followed any physical cause. Ascertain whether it is a first attack, whether the patient has suffered from epilepsy, has squandered his money, grown restless, has absurd delusions, etc. In order to ascertain the capacity of the mind, questions should be asked with regard to age, birthplace, profession, number of family, and common events, such as the day of week, month, and year. The power of performing simple arithmetical operations may be tested. It may be necessary to pay more than one visit. The examiner should be careful to ask questions adapted to the station of life of the supposed lunatic; a man is not necessarily mad because he cannot perform simple arithmetical operations, or does not know about things with which his questioner is well acquainted. The opinion of a supposed lunatic that his examiner's feet were large was not considered by the Commissioners among the facts indicating insanity, yet statements quite as absurd are made by medical men as 'facts of insanity' observed by themselves. 'Reads his Bible and is anxious about the salvation of his soul' is another example of a bad certificate. Some well-marked delusion should be recorded.

For a lunacy certificate (Reception Order on Petition or Judicial Reception Order), except in the case of a pauper patient, there are required the signatures of two independent medical men and of a relation or friend. The medical men must not be in partnership or in any way interested in the patient; they must make separate visits at different times, and write on the proper forms the facts observed by themselves and those observed by others, giving the name of the informer. A certificate is valid only for seven days. In very urgent non-pauper cases the signature of one medical man is sufficient, but such certificate (Emergency Certificate or Urgency Order) is only valid for two days, and, as the patient can only be detained in the asylum under this order for seven days in England or three in Scotland, it must be supplemented by another signed as above directed. The medical certificate must contain a statement that it is expedient for the alleged lunatic to be placed forthwith under care, with reasons for making such statement. The certifying medical practitioner must have personally examined the patient not more than two clear days before his reception. In London and other large towns, where an expert opinion is readily obtainable, it is not expedient to resort to such urgency orders. Medical men should be careful how they sign certificates of insanity. No medical man is bound to certify, but if he does so he must be prepared to take the responsibility of his acts. There must be no reasonable ground for alleging want of 'good faith' or 'reasonable care.' The practitioner must exercise that amount of care and skill which he may reasonably be expected to possess.

XLV

THE INEBRIATES ACTS

It is somewhat difficult to define an inebriate, but for the moment the following will suffice, and will ultimately, in all probability, be officially adopted:

An inebriate is a person who habitually takes or uses any intoxicating thing or things, and while under the influence of such thing or things, or in consequence of the effects thereof, is—(a) dangerous to himself or others; or (b) a cause of harm or serious annoyance to his family or others; or (c) incapable of managing himself or his affairs, or of ordinary proper conduct.

Under the provisions of the Habitual Drunkards Acts (42 and 43 Vict., c. 19, and 51 and 52 Vict., c. 19), any habitual drunkard may voluntarily place himself under restraint. He must make an application to the owner of a licensed retreat, stating the time during which he undertakes to remain. His application must be accompanied by a statutory declaration of two persons stating that they knew the applicant to be a confirmed drunkard. Without this testimony as to moral character his application cannot be entertained. His signature must also be attested by two justices, who must state that he understands the effect of his application, and that it has been explained to him. The limit to the term of restraint is twelve months, after which he must resume his former habits if he wishes to qualify for another period. The Act works automatically, and, when it has been set for a certain time, the patient cannot release himself until the period has expired. The Inebriates' Retreat must be duly licensed, and the licensee incurs distinct obligation in return for the powers entrusted to him. It is an offence against the Act to assist any habitual drunkard to escape from his retreat, and should he succeed in effecting his escape he may be arrested on a warrant. A drunkard who does not obey orders and conform to the rules of the establishment may be sent to prison for seven days. It may be as well to mention that it is an offence to supply any drunkard under the Act with any intoxicating drink or sedative or stimulant drug without authority, and that the penalty is a fine of £20 or three months' imprisonment. The Act is a good one, but might be carried farther with advantage. It has been ruled that a crime committed during drunkenness is as much a crime as if committed during sobriety. A person is supposed to know the effect of drink, and if he takes away his senses by drink it is no excuse. He is held answerable both for being under the influence of alcohol or of any other drug, and for the acts such influence induces.

Inebriates Act (1898-1900).—If an habitual drunkard be sentenced to imprisonment or penal servitude for an offence committed during drunkenness, or if he has been convicted four times in one year, the court may order him to be detained for a term not exceeding three years in an inebriate reformatory.

PART II

TOXICOLOGY

I

DEFINITION OF A POISON

Though the law does not define in definite terms what a poison really is, it lays stress on the malicious intention in giving a drug or other substance to an individual. It is a felony to administer, or cause to be administered, any poison or other destructive thing with intent to murder, or with the intention of stupefying or overpowering an individual so that any indictable offence may be committed. It is a misdemeanour to administer any poison, or destructive or noxious thing, merely to aggrieve, injure, or annoy an individual. For a working definition we may state that a poison is a substance which, when introduced into or applied to the body, is capable of injuring health or destroying life. A poison may therefore be swallowed, applied to the skin, injected into the tissues, or introduced into any orifice of the body.

II

SALE OF POISONS; SCHEDULED POISONS

The sale of poisons is regulated by various Acts, but chiefly by the Pharmacy Act, 1868, and by the Poisons and Pharmacy Act, 1908.

Only registered medical practitioners and legally qualified druggists are permitted to dispense and sell scheduled poisons. They are responsible for any errors which may be committed in the sale of poisons. If a druggist knows that a drug in a prescription is to be used for an improper purpose, he may refuse to dispense it. The practitioner who carelessly prescribes a drug in a poisonous dose is not held responsible, but the dispenser would be if he dispensed it and harmful or fatal consequences followed on its being swallowed. When a dispenser finds an error in a prescription, it is his duty to communicate with the prescriber privately pointing out the mistake.

A great responsibility rests on the medical man who does his own dispensing, as there is no one to check his work.

If a doctor prescribes a drug with the intention of curing or preventing a disease, but that, contrary to expectation and general experience, it causes illness or even death, no responsibility can rest with the prescriber. It has to be proved that actual injury has been sustained by the complainant before an action for damages can be commenced, and that the plaintiff was free from all contributory negligence.

Scheduled Poisons.—By the Pharmacy Act of 1868 two groups of poisons are scheduled. Part I. contains a list of those which are considered very active poisons—e.g., arsenic, alkaloids, belladonna, cantharides, coca (if containing more than 1 per cent. alkaloids), corrosive sublimate, diachylon, cyanides, tartar emetic, ergot, nux vomica, laudanum, opium, savin, picrotoxin, veronal and all poisonous urethanes, prussic acid, vermin killers, etc. Such poisons must not be sold to strangers, but only to persons known to or introduced by someone known to the druggist. If sold, the latter must enter into the 'Poison Register' the name of the poison, the name of the person to whom it is sold, the quantity and purpose for which it is to be used, and date of sale. The entry must be signed by the purchaser and by the introducer. The word 'Poison' must be affixed to the bottle or package, and also the name and address of the seller.

Part II. contains a list of poisons supposed to be less active. These may only be sold if on the bottle, box, or package there is affixed a label with the name of the article, the word 'Poison,' and the name and address of the seller. It is not necessary to enter the transaction in a register.

Chemists are required to keep poisons in specially distinguishable bottles, and these in a special room or locked cupboard.

Dangerous Drugs Act, 1920.—The regulations restrict the manufacture and sale of opium, morphine, cocaine, and heroin so as to prevent their abuse. Preparations containing less than 1/5 per cent. of the first two or less than 1/10 per cent. of the last two are excluded. Prescriptions containing the above drugs must be dated and signed with the full name and address of the prescriber, and must have also those of the patient. The total amount of the drug to be supplied must be stated, and it must not be dispensed more than once; the dispenser retains the prescription. Special books must be kept recording the purchase and sale of these drugs.

Proprietary Medicines Bill (introduced in 1920, and likely soon to become law).—The sale of any unregistered proprietary medicine purporting to cure certain diseases or produce abortion is made an offence. A register of proprietary medicines, etc., is established. The object is to protect the public against quack remedies.

Notification of Poisoning.—Every case of poisoning which occurs in any industry (lead, arsenic, anthrax, etc.) must be notified by the medical attendant to the Chief Inspector of Factories (Factory and Workshops Act, 1895).

III

ACTION OF POISONS; CLASSIFICATION OF POISONS

Action of Poisons.—They may act either locally or only after absorption into the system.

1. *Local Action*, as seen in (a) corrosive poisons; (b) irritant poisons, causing congestion and inflammation of the mucous membranes—e.g., metallic and vegetable irritants; (c) stimulants or sedatives to the nerve endings, as aconite, conium, cocaine.

2. *Remote Action.*—This may be of reflex character, as seen in the shock produced by the pain caused by corrosive poisons, or the poison may exert a special action on certain structures, as belladonna on the cells of the brain, strychnine on the motor nerve cells of the spinal cord.

3. *In Both Ways.*—Certain poisons, as carbolic or oxalic acids, act in this way.

Age, idiosyncrasy, tolerance, and disease, all exert modifying influences on the action of a poison. The form in which the poison is swallowed and the quantity also determine its action. In the gaseous form, poisons act most rapidly and fatally. When in solution and injected hypodermically, they also act very rapidly. In the solid form they act as a rule slowly, and may even set up vomiting, and so may be entirely ejected by vomiting. Poisons act most energetically when the stomach is empty. If taken when the stomach already contains food, solution and absorption may be greatly delayed.

Some poisons are cumulative in their action, and thus, even if infinitesimal doses be swallowed each day, there is a certain amount of storage in the tissues (though a certain percentage of the poison is being constantly eliminated), and at last symptoms of poisoning show themselves.

Classification of Poisons.—As an aid to memory, the following classification is perhaps the best:

I Inorganic.

1. Corrosive acids and alkalies, and caustic salts (carbolic and oxalic acids also).

2. Irritant—practically all the metals and the metalloids (I. Cl. Br. P.).

II Organic.

1. Irritant
   - Animal—venomous bites, food poisoning, cantharides.
   - Vegetable—all strong purgatives, hellebores, savin, yew, ergot, hemlock, laburnum, bryony, etc.

2. Neuronic.

   (a) Somniferous—opium and its alkaloids.
   (b) Deliriant—belladonna, hyoscyamus, stramonium, cannabis, cocaine, cocculus, camphor, fungi.
   (c) Inebriants—alcohol, ether, chloral, carbolic acid (weak), benzol, aniline, nitro-glycerine.

3. Sedative or depressant.

   (a) Neural—conium, lobelia, tobacco, physostigma.
   (b) Cerebral—hydrocyanic acid.
   (c) Cardiac—aconite, digitalis, colchicum, veratrum.

4. Excito-motory or convulsives—nux vomica, strychnine.

5. Vulnerants—powdered glass.

III Asphyxiants.

Poisonous and irrespirable gases.

IV

EVIDENCE OF POISONING

It may be inferred that poison has been taken from consideration of the following factors: Symptoms and post-mortem appearances, experiments on animals, chemical analysis, and the conduct of suspected persons.

1. *Symptoms* in poisoning usually come on suddenly, when the patient is in good health, and soon after taking a meal, drink, or medicine. Many diseases, however, come on suddenly, and in cases of slow poisoning the invasion of the symptoms may be gradual.

2. *Post-Mortem Appearances.*—These in many poisons and classes of poisons are characteristic and unmistakable. The post-mortem appearances peculiar to the various poisons will be described in due course.

3. *Experiments on Animals.*—These may be of value, but are not always conclusive.

4. *Chemical Analysis.*—This is one of the most important forms of evidence, as a demonstration of the actual presence of a poison in the body carries immense weight. The poison may be discovered in the living person by testing the urine, the blood abstracted by bleeding, or the serum of a blister. In the dead body it may be found in the blood, muscles, viscera—especially the liver—and secretions. Its discovery in these cases must be taken as conclusive evidence of administration. If, however, it be found only in

substances rejected or voided from the body, the evidence is not so conclusive, as it may be contended that the poison was introduced into or formed in the material examined after its rejection from the body, or if the quantity be very minute it will be argued that it is not sufficient to cause death. A poison may not be detected in the body, owing to defective methods, smallness of the dose required to cause death, or to its ejection by vomiting or its elimination by the excretions.

5. *Conduct of Suspected Persons.*—A prisoner may be proved to have purchased poison, to have made a study of the properties and effects of poison, to have concocted medicines or prepared food for the deceased, to have made himself the sole attendant of the deceased, to have placed obstacles in the way of obtaining proper medical assistance, or to have removed substances which might have been examined.

V

SYMPTOMS AND POST-MORTEM APPEARANCES OF DIFFERENT CLASSES OF POISONS

Whilst recognizing the fact that toxic agents cannot be accurately classified, the following grouping may for descriptive purposes be admitted with the view of saving needless repetition:

1. Corrosives.—Characterized by their destructive action on tissues with which they come in contact. The principal inorganic corrosives are the mineral acids, the caustic alkalies, and their carbonates; the organic are carbolic acid, strong solutions of oxalic acid, and acetic acid.

Symptoms.—Burning pain in mouth, throat, and gullet, strong acid, metallic or alkaline taste; retching and vomiting, the discharged matters containing shreds of mucus, blood, and the lining membrane of the passages. Inside of mouth corroded. There are also dysphagia, thirst, dyspnœa, small and frequent pulse, anxious expression, shock. Death may result from shock, destruction of the parts—e.g., perforation of stomach or duodenum, suffocation; or some weeks subsequently death may be due to cicatricial contraction of the gullet, stomach, or pylorus.

Post-Mortem Appearances.—Those of corrosion, with corrugation from strong contraction of muscular fibres, and followed by inflammation and its consequences. The mouth, gullet, and stomach, and in some cases the intestines, may be white, yellow, or brown, shrivelled and corroded. The corrosions may be small, or may extend over a very large surface. Sometimes considerable portions of the lining membrane of the gullet or stomach may be discharged by vomiting or by stool. Beyond the corroded parts the textures are acutely inflamed. The stomach is filled with a yellow, brown, or black gelatinous liquid or black blood, and may in rare cases be perforated.

2. Irritants.—These are substances which inflame parts to which they are applied. The class includes mineral, animal, and vegetable substances, and contains a larger number of poisons than all the other classes together. Irritants may be divided into two groups: (1) Those which destroy life by the irritation they set up in the parts to which they are applied; (2) those which add to local irritation peculiar or specific remote effects. The first group includes the principal vegetable irritants, some alkaline salts, some metallic poisons, etc.; and the second comprises the metallic irritants, the metalloids (phosphorus and iodine), and one animal substance, cantharides.

Symptoms.—Burning pain and constriction in throat and gullet, pain and tenderness of stomach and bowels, intense thirst, nausea, vomiting, purging and tenesmus, with bloody stools, dysuria, cold skin, and feeble and irregular pulse. The vomit consists at first of the food, then it becomes bile-stained, and later dark coffee-grounds in appearance, due to extravasation of blood from the over-distended vessels in the gastric mucous membrane. Death may occur from shock, convulsions, collapse, exhaustion, or from starvation on account of chronic inflammation of the gastro-intestinal mucous membrane.

Post-Mortem Appearances.—Those of inflammation and its consequences. Coats of stomach, fauces, gullet, and duodenum, may be thickened, black, ulcerated, gangrenous, or sloughing. Vessels filled with dark blood ramify over the surface. Acute inflammation is often found in the small intestines, with ulceration and softening of mucous membrane. The rectum is frequently the seat of marked ulceration.

3. Poisons Acting on the Brain.—Three classes: The opium group, producing sleep; the belladonna group, producing delirium and illusions; and the alcohol group, causing exhilaration, followed by delirium or sleep.

Symptoms.—Of the opium group, giddiness, headache, dimness of sight, contraction of the pupils, noises in the ears, drowsiness and confusion, passing into insensibility. Of the belladonna group, delirium, illusions of sight, dilated pupils, dry mouth, thirst, redness of skin, coma. Of the alcohol group, excitement of circulation and of cerebral functions, want of power of co-ordination and of muscular movement, double vision, mania, followed by profound sleep and coma. In the chronic form, delirium tremens.

Post-Mortem Appearances.—In the opium group, fulness of the sinuses and veins of the brain, with effusion of serum into the ventricles and beneath the membranes. In the belladonna group, nil. In the alcohol group, signs of inflammation, congestion of brain and membranes, fluidity of blood, long-continued rigor mortis.

4. Poisons Acting on the Spinal Cord.—Strychnine, brucine, thebaïne. The leading symptom is tetanic spasm.

5. Poisons Affecting the Heart.—These kill by sudden shock, syncope, or collapse. They comprise prussic acid, dilute solution of oxalic acid and oxalates, aconite, digitalis, strophanthus, convallaria, and tobacco.

6. Poisons Acting on the Lungs.—These have for their type carbonic acid gas and coal gas. The fumes of ammonia are intensely irritating, and may give rise to laryngitis, bronchitis, and even pneumonia. Nitric acid fumes sometimes produce no serious symptoms for an hour or more, but there may then be coughing, difficulty of breathing, and tightness in the lower part of the throat, followed by capillary bronchitis (see p. 124).

VI

DUTY OF PRACTITIONER IN SUPPOSED CASE OF POISONING

If called to a case supposed or suspected to be one of poisoning, the medical man has two duties to perform: To save the patient's life, and to place himself in a position to give evidence if called on to do so. If life is extinct, his duty is a simple one. He should make inquiries as to symptoms, and time at which food or medicine was last taken. He should take possession of any food, medicine, vomited matter, urine, or fæces, in the room, and should seal them up in clean vessels for examination. He should notice the position and temperature of the body, the condition of rigor mortis, marks of violence, appearance of lips and mouth. He should not make a post-mortem examination without an order in writing from the coroner. In making a post-mortem examination, the alimentary canal should be removed and preserved for further investigation. A double ligature should be passed round the œsophagus, and also round the duodenum a few inches below the pylorus. The gut and the gullet being cut across between these ligatures, the stomach may be removed entire without spilling its contents. The intestines may be removed in a similar way, and the whole or a portion of the liver should be preserved. These should all be put in separate jars without any preservative fluid, tied up, sealed, labelled, and initialled. All observations should be at once committed to writing, or they will not be admitted by the court for the purpose of refreshing the memory whilst giving evidence. If the medical practitioner is in doubt on any point, he should obtain technical assistance from someone who has paid attention to the subject.

In a case of attempted suicide by poisoning, is it the duty of the doctor to inform the police? He would be unwise to do so. He had much better stick to his own business, and not act as an amateur detective.

VII

TREATMENT OF POISONING

The modes of treatment may be ranged under three heads: (1) To eliminate the poison; (2) to antagonize its action; (3) to avert the tendency to death.

1. The first indication is met by the administration of emetics, to produce vomiting, or by the application of the stomach-tube. The best emetic is that which is at hand. If there is a choice, give apomorphine hypodermically. The dose for an adult is 10 minims. It may be given in the form of the injection of the Pharmacopœia, or preferably as a tablet dissolved in water. Apomorphine is not allied in physiological action to morphine, and may be given in cases of narcotic poisoning. Sulphate of zinc, salt-and-water, ipecacuanha, and mustard, are all useful as emetics. Tickling the fauces with a feather may excite vomiting.

In using the elastic stomach-tube, some fluid should be introduced into the stomach before attempting to empty it, or a portion of the mucous membrane may be sucked into the aperture. The tube should be examined to see that it is not broken or cracked, as accidents have happened from neglecting this precaution. The bowels and kidneys must also be stimulated to activity, to help in the elimination of the poison.

2. The second indication is met by the administration of the appropriate antidote. Antidotes are usually given hypodermically, or, if by mouth, in the form of tablets. In the absence of a hypodermic syringe, the remedy may be given by the rectum. In the selection of the appropriate antidote, a knowledge of pharmacology is required, especially of the physiological antagonism of drugs. Antidotes may act (1) chemically, by forming harmless compounds, as lime in oxalic acid poisoning; (2) physiologically, the drug which is administered neutralizing more or less completely the poison which has been absorbed; (3) physically, as charcoal. Every doctor should provide himself with an antidote case. The various antidotes will be mentioned under their respective poisons.

3. To avert the tendency to death, we must endeavour to palliate the symptoms and neutralize the effects of the poison. Pain must be relieved by the use of morphine; inflamed mucous membrane soothed by such demulcents as oils, milk, starch; stimulants to overcome collapse; saline infusions in shock, etc. In the case of narcotics and depressing agents, stimulants, electricity, and cold affusions, may be found useful. We should endeavour to promote the elimination of the poison from the body by stimulating the secretions.

VIII

DETECTION OF POISONS

Notice the smell, colour, and general appearance, of the matter submitted for examination. The odour may show the presence of prussic acid, alcohol, opium, or phosphorus. The colour may indicate salts of copper, cantharides, etc. Seeds of plants may be found.

This examination having been made, the contents of the alimentary canal, and any other substances to be examined, must be submitted to chemical processes.

Simple filtration will sometimes suffice to separate the required substance; in other cases dialysis will be necessary, in order that crystalloid substances may be separated from colloid bodies.

In the case of volatile substances distillation will be required. The poisons thus sought for are alcohol, phosphorus, iodine, chloral, ether, hydrocyanic acid, carbolic acid, nitro-benzol, chloroform, and anilin. The organic matters are placed in a flask, diluted with distilled water if necessary, and acidulated with tartaric acid. The flask is heated in a water-bath, and the vapours condensed by a Liebig's condenser. In the case of phosphorus the condenser should be of glass, and the process of distillation conducted in the dark, so that the luminosity of the phosphorus may be noted.

For the separation of an alkaloid, the following is the process of Stas-Otto. This process is based upon the principle that the salts of the alkaloids are soluble in alcohol and water, and insoluble in ether. The pure alkaloids, with the exception of morphine in its crystalline form, are soluble in ether. Make a solution of the contents of the stomach or solid organs minced very fine by digesting them with acidulated alcohol or water and filtering. The filtrate is shaken with ether to remove fat, etc., the ether separated, the watery solution neutralized with soda, and then shaken with ether, which removes the alkaloid in a more or less impure condition. The knowledge of these facts will help to explain the following details, which may be modified to suit individual cases: (1) Treat the organic matter, after distillation for the volatile substances just mentioned, with twice its weight of absolute alcohol, free from fusel oil, to which from 10 to 30 grains of tartaric or oxalic acid have been added, and subject to a gentle heat. (2) Cool the mixture and filter; wash the residue with strong alcohol, and mix the filtrates. The residue may be set aside for the detection of the metallic poisons, if suspected. Expel the alcohol by careful evaporation. On the evaporation of the alcohol the resinous and fatty matters separate. Filter through a filter moistened with water. Evaporate the filtrate to a syrup, and extract with successive portions of absolute alcohol. Filter through a filter moistened with alcohol. Evaporate filtrate to dryness, and dissolve residue in water, the solution being made distinctly acid. Now shake watery solution with ether. (3) Ether from the acid solution dissolves out colchicin, digitalin, cantharidin, and picrotoxin, and traces of veratrine and atropine. Separate the ethereal solution and evaporate. Hot water will now dissolve out picrotoxin, colchicin, and digitalin, but not cantharidin. (4) The remaining acid watery liquid, holding the other alkaloids in solution or suspension, is made strongly alkaline with soda, mixed with four or five times its bulk of ether, chloroform, or benzole, briskly shaken, and left to rest. The ether floats on the surface, holding the alkaloids, except morphine, in solution. (5) A part of this ethereal solution is poured into a watch-glass and allowed to evaporate. If the alkaloid is volatile, oily streaks appear on the glass; if not volatile, crystalline traces will be visible. If a volatile alkaloid, add a few pieces of calcium chloride to ethereal solution to absorb the water; draw off the ethereal solution with a pipette, allow it to evaporate, and test the residue for the alkaloids, conine and nicotine.

If a fixed alkaloid, treat the acid solution with soda or potash and ether, evaporate ethereal solution after separation, when the solid alkaloid will be left in an impure state. To purify it, add a small quantity of dilute sulphuric acid, and, after evaporating to three-quarters of its bulk, add a saturated solution of carbonate of potash or soda. Absolute alcohol will then dissolve out the alkaloid, and leave it on evaporation in a crystalline form.

General Reactions for Alkaloids.—(1) Wagner's reagent (iodine dissolved in a solution of potassium iodide) yields a reddish-brown precipitate; (2) Mayer's reagent (potassio-mercuric iodide) gives a yellowish-white precipitate; (3) phospho-molybdic acid gives a yellow precipitate; (4) platinic chloride, a brown precipitate; (5) tannic acid, etc.

In order to isolate an inorganic substance from organic matter, Fresenius's method is adopted. Boil the finely divided substance with about one-eighth its bulk of pure hydrochloric acid; add from time to time potassic chlorate until the solids are reduced to a straw-yellow fluid. Treat this with excess of bisulphate of sodium, then saturate with sulphuretted hydrogen until metals are thrown down as sulphides. These may be collected and tested. From the acid solution, hydrogen sulphide precipitates copper, lead, and mercury, dark; arsenic, antimony, and tin, yellowish. If no precipitate, add ammonia and ammonium sulphide, iron, black, zinc, white, chromium, green, manganese, pink. The residue of the material after digestion with hydrochloric acid and potassium chlorate may have to be examined for silver, lead, and barium.

For the detection of minute quantities, the microscope must be used, and Guy's and Helwig's method of sublimation will be found advantageous. Crystalline poisons may be recognized by their characteristic forms.

IX

THE MINERAL ACIDS

These are sulphuric, nitric, and hydrochloric acids.

Symptoms of Poisoning by the Mineral Acids.—Acid taste in the mouth, with violent burning pain extending into the œsophagus and stomach, and commencing immediately on the poison being swallowed; eructations, constant retching, and vomiting of brown, black, or yellow matter containing blood, coagulated mucus, epithelium, or portions of the lining membrane of the gullet and stomach. The vomited matters are strongly acid in reaction, and stain articles of clothing on which they may fall. There is intense thirst and constipation, with scanty or suppressed urine, tenesmus, and small and frequent pulse; the lips, tongue, and inside of the mouth, are shrivelled and corroded. Exhaustion succeeds, and the patient dies either collapsed, convulsed, or suffocated, the intellect remaining clear to the last. After recovering from the acute form of poisoning, the patient may ultimately die from starvation, due to stricture of the œsophagus, stomach, etc.

Post-Mortem Appearances Common to the Mineral Acids.—Stains and corrosions about the mouth, chin, and fingers, or wherever the acid has come in contact. The inside of the mouth, fauces, and œsophagus, is white and corroded, yellow or dark brown, and shrivelled. Epiglottis contracted or swollen. Stomach filled with brown, yellow, or black glutinous liquid; its lining membrane is charred or inflamed, and the vessels are injected. Pylorus contracted. Perforation, when it takes place, is on the posterior aspect; the apertures are circular, and surrounded by inflammation and black extravasation. The blood in the large vessels may be coagulated.

Avoid mistaking gastric or duodenal ulcer, with or without perforation, for the effects of a corrosive poison.

Treatment.—Calcined magnesia or the carbonate or bicarbonate of sodium, mixed with milk or some mucilaginous liquid, are the best antidotes. In the absence of these, chalk, whiting, milk, oil, soap-suds, etc., will be found of service. The stomach-pump should not be used. If the breathing is impeded, tracheotomy may be necessary. Injuries of external parts by the acid must be treated as burns.

X

SULPHURIC ACID

Sulphuric Acid, or oil of vitriol, may be concentrated or diluted. It is frequently thrown over the person to disfigure the features or destroy the clothes. Parts of the body touched by it are stained, first white, and then dark brown or black. The presence of corrosion of the mouth is as important as the chemical tests. Black woollen cloths are turned to a dirty brown, the edges of the spots becoming red in a few days, due to the dilution of the acid from the absorption of moisture; the stains remain damp for long, owing to the hygroscopic property of the acid.

Method of Extraction from the Stomach.—The contents of the stomach or vomited matter should, if necessary, be diluted with pure distilled water and filtered. The stomach should be cut up into small pieces and boiled for some time in water. The solution, filtered and concentrated, is now ready for testing. Blood, milk, etc., may be separated by dialysis, and the fluid so obtained tested. A sulphate may be present. Take a portion of the liquid, evaporate to dryness, and incinerate; a sulphate, if present, will be obtained, and may be tested.

Caution.—Sulphuric acid may not be found even after large doses, due to treatment, vomiting, or survival for several days. In all cases every organ should be examined. Vomited matters and contents of stomach should not be mixed, but each separately examined. This rule holds good for all poisons. On cloth the stain may be cut out, boiled in water, the solution filtered, and tested with blue litmus and other tests.

Post-Mortem Appearances.—Where the acid has come in contact with the mucous membranes there are dark brown or black patches. The stomach is greatly contracted, the summits of the mucous membrane ridges being charred and the furrows greatly inflamed; the contents are black or brown.

Tests.—Concentrated acid chars organic matter; evolves heat when added to water, and sulphurous fumes when boiled with chips of wood, copper cuttings, or mercury. Dilute acid chars paper when the paper is heated; gives a white precipitate with nitrate or chloride of barium, and is entirely volatilized by heat. Dilute solutions give a white precipitate with barium nitrate, insoluble in hydrochloric acid even on boiling.

Fatal Dose.—In an adult, 1 drachm.

Fatal Period.—Shortest, three-quarters of an hour; average period from onset of primary effects, eighteen to twenty-four hours.

XI

NITRIC ACID

Nitric Acid, or aqua fortis, is less frequently used as a poison than sulphuric acid. The fumes from nitric acid have caused death from pneumonia in ten or twelve hours.

Method of Extraction from the Stomach.—The same as for sulphuric acid. In beer, etc., the mixture may be neutralized with carbonate of potassium, dialyzed, the fluid concentrated and allowed to crystallize, when crystals of nitrate of potassium may be recognized.

Post-Mortem Appearance.—The mucous membranes are rendered yellow or greenish if bile be present; they are also thickened and hardened.

Tests.—Concentrated acid gives off irritating orange-coloured fumes of nitric acid gas. When poured on copper, it gives off red fumes and leaves a green solution of nitrate of copper. It gives a red colour with brucine, turns the green sulphate of iron black, and with hydrochloric acid dissolves gold. A delicate test for the acid, free or in combination, is to dissolve in the suspected fluid some crystals of ferrous sulphate, and then to gently pour down the test-tube some strong sulphuric acid. Where the two liquids meet, if nitric acid be present, a reddish-brown ring will be formed. It turns the skin bright yellow, and does the same with woollen clothes, from the formation of picric acid.

Fatal Dose.—Two drachms.

Fatal Period.—Shortest, one hour and three-quarters in an adult; in infants in a few minutes, from suffocation.

XII

HYDROCHLORIC ACID

Hydrochloric Acid, muriatic acid, or spirit of salt, is not uncommonly used for suicidal purposes, being fifth in the list.

Method of Extraction from the Stomach.—The same as for sulphuric acid. As hydrochloric acid is a constituent of the gastric juice, the signs of the acid must be looked for.

Post-Mortem Appearances.—The mucous membranes are dry, white, and shrivelled, and often eroded.

Tests.—The concentrated acid yields dense white fumes with ammonia. When warmed with black oxide of manganese and strong sulphuric acid it gives off chlorine, recognized by its smell and bleaching properties. Diluted it gives with nitrate of silver, a white precipitate, which is insoluble in nitric acid and in caustic potash, but is soluble in ammonia, and when dried and heated melts, and forms a horny mass. Stains on clothing are reddish-brown in colour.

Fatal Dose.—Half an ounce.

Fatal Period.—Shortest, two hours; average, twenty-four hours. Death may occur after an interval of some weeks from destruction of the gastric glands and inability to digest food.

XIII

OXALIC ACID

Oxalic Acid is used by suicides, though not often by murderers. The crystals closely resemble those of Epsom salts or sulphate of zinc; oxalic acid has been taken in mistake for the former. It is in common use for cleansing brass, in laundry work, for dyeing purposes, and especially for bleaching straw hats.

Symptoms.—If a concentrated solution be taken, it acts as a corrosive, causing a burning acid, intensely sour taste, which comes on immediately, great pain and tenderness and burning at pit of stomach, pain and tightness in throat. Vomiting of mucus, bloody or dark coffee-ground matters, purging and tenesmus, followed by collapse, feeble pulse, cyanosis and pallor of the skin; also swelling of tongue, with dysphagia. In some cases cramps and numbness in limbs, pain in head and back, delirium and convulsions. May be tetanus or coma. If taken freely diluted, the nervous symptoms predominate, and may resemble narcotic poisoning. Sometimes almost instant death.

Post-Mortem Appearances.—Mucous membrane of mouth, throat, and gullet, white and softened, as if they had been boiled; there are often black or brown streaks in it. Stomach contains dark, grumous matter, and is soft, pale, and brittle. Intestines slightly inflamed, stomach sometimes quite healthy.

Treatment.—Warm water, then chalk, carbonate of magnesium, or lime-water, freely. Not alkalies, as the oxalates of the alkalies are soluble and poisonous. Castor-oil. Emetics, but not stomach-pump.

Fatal Dose.—One drachm is the smallest, but half an ounce is usually fatal.

Method of Extraction from the Stomach.—Mince up the coats of the stomach and boil them in water, or boil the contents of the stomach and subject them to dialysis. Concentrate the distilled water outside the tube containing the vomited matters, etc., and apply tests.

Tests.—White precipitate with nitrate of silver, soluble in nitric acid and ammonia. When the precipitate is dried and heated on platinum-foil, it disperses as white vapour with slight detonation. Sulphate of lime in excess gives a white precipitate, soluble in nitric or hydrochloric acid, but insoluble in oxalic, tartaric, acetic, or any vegetable acid.

Oxalate or Binoxalate of Potash (salts of sorrel or salts of lemon) is almost as poisonous as the acid itself.

XIV

CARBOLIC ACID

Carbolic Acid, Phenic Acid, or Phenol, is largely employed as a disinfectant, and is often supplied in ordinary beer-bottles without labels.

Symptoms.—An intense burning pain extending from the mouth to the stomach and intestines. Indications of collapse soon supervene. The skin is cold and clammy, and the lips, eyelids, and ears, are livid. This is followed by insensibility, coma, stertorous breathing, abolition of reflex movements, hurried and shallowed respiration, and death. The pupils are usually contracted, and the urine, if not suppressed, is dark in colour, or even black. Patients often improve for a time, and then die suddenly from collapse. When the poison has been absorbed through the skin or mucous membranes, a mild form of delirium, with great weakness and lividity, are the first signs.

Post-Mortem.—If strong acid has been swallowed, the lips and mucous membranes are hardened, whitened, and corrugated. In the stomach the tops of the folds are whitened and eroded, while the furrows are intensely inflamed.

Treatment.—Soluble sulphates which form harmless sulpho-carbolates in the blood should be administered at once. An ounce of Epsom salts or of Glauber's salts dissolved in a pint of water will answer the purpose admirably. After this an emetic of sulphate of zinc may be given. White of egg and water or olive-oil may prove useful. Warmth should be applied to the body.

Fatal Dose.—One drachm, but recovery has taken place after much larger quantities, if well diluted or taken after a meal.

Tests are not necessary, as the smell of carbolic acid is characteristic.

Local action of carbolic acid produces anæsthesia and necrosis. Accidents sometimes happen from too strong lotions applied as surgical dressings.

Lysol is a compound of cresol and linseed-oil soap, and is much less toxic than carbolic acid.

XV

POTASH, SODA, AND AMMONIA

Caustic Potash occurs in cylindrical sticks, is soapy to the touch, has an acrid taste, is deliquescent, fusible by heat, soluble in water. Liquor Potassæ is a strong solution of caustic potash, and has a similar reaction. Carbonate of Potassium, also known as potash, pearlash, salt of tartar, is a white crystalline powder, alkaline and caustic in taste, and very deliquescent. The bicarbonate is in colourless prisms, which have a saline, feebly alkaline taste, and are not deliquescent.

Symptoms.—Acrid soapy taste in mouth, burning in throat and gullet, acute pain at pit of stomach, vomiting of bloody or brown mucus, colicky pains, bloody stools, surface cold, pulse weak. These preparations are not volatile, so that there is not much fear of lung trouble. In chronic cases death occurs from stricture of the œsophagus causing starvation.

Post-Mortem Appearances.—Soapy feeling, softening, inflammation, and corrosion of mucous membrane of mouth, pharynx, œsophagus, stomach, and intestines. Inflammation may have extended to larynx.

Method of Extraction from the Stomach.—If the contents of the stomach have a strong alkaline action, dilute with water, filter, and apply tests.

Tests.—The carbonates effervesce with an acid. The salts give a yellow precipitate with platinum chloride, and a white precipitate with tartaric acid. They are not dissipated by heat, and give a violet colour to the deoxidizing flame of the blowpipe. Stains on dark clothing are red or brown.

Treatment.—Vinegar and water, lemon-juice and water, acidulated stimulant drinks, oil, linseed-tea, opium to relieve pain, stimulants in collapse. Do not use the stomach-tube. The glottis may be inflamed, and if there is danger of asphyxia, tracheotomy may have to be performed.

Carbonate of Sodium occurs as soda and best soda, the former in dirty crystalline masses, the latter of a purer white colour. It is also found as 'washing soda.'

Symptoms, Post-Mortem Appearances, Treatment, and Extraction from the Stomach.—As for potash.

Tests.—Alkaline reaction, effervesces and evolves carbonic acid when treated with an acid; crystallizes, gives yellow tinge to blowpipe flame. No precipitate with tartaric acid, nor with bichloride of platinum.

Ammonia may be taken as liquor ammoniæ (harts-horn), as carbonate of ammonium, as 'Cleansel,' or as 'Scrubb's Cloudy Ammonia.'

Symptoms.—Being volatile, it attacks the air-passages, nose, eyes and lungs, being immediately affected; profuse salivation; lips and tongue swollen, red, and glazed. The urgent symptoms are those of suffocation.

Inhalation of the fumes of strong ammonia may lead to death from capillary bronchitis or broncho-pneumonia. Death may result from inflammation of the larynx and lungs. When swallowed in solution, the symptoms are similar to those of soda and potash.

Post-Mortem Appearances.—Similar to other corrosives.

Method of Extraction from the Stomach.—The contents of the stomach, etc., must be first distilled, the gas being conveyed into water free from ammonia.

Tests.—Nessler's reagent is the most delicate, a reddish-brown colour or precipitate being produced, but ammonia may be recognized by its pungent odour, dense fumes given off with hydrochloric acid, and strong alkaline reaction.

Treatment.—Vinegar and water. Other treatment according to symptoms.

Fatal Dose.—One drachm of strong solution.

Fatal Period (Shortest).—Four minutes.

XVI

INORGANIC IRRITANTS

Nitrate of Potassium (Nitre, Saltpetre)—Bitartrate of Potassium (Cream of Tartar)—Alum (Double Sulphate of Alumina and Potassium)—Chlorides of Lime, Sodium, and Potassium.—All these are irritant drugs, and give the usual symptoms.

XVII

CHLORATE OF POTASSIUM, ETC.

Chlorate of Potassium produces irritation of stomach and bowels; hæmaturia; melæna; cyanosis, weakness, delirium, and coma.

Post-Mortem.—Blood is chocolate-brown in colour, and so are all the internal organs; gastro-enteritis; nephritis.

Tests.—Spectroscope shows blood contains methæmoglobin; the drug discharges the colour of indigo in acid solution with SO_2.

Treatment.—Transfusion of blood or saline fluid; stimulants.

Sulphuret of Potassium (liver of sulphur) occurs in mass or powder of a dirty green colour; has a strong smell of sulphuretted hydrogen.

Symptoms.—Of acute irritant poisoning, with stupor or convulsions. Excreta smell of sulphuretted hydrogen.

Post-Mortem Appearances.—Stomach and duodenum reddened, with deposits of sulphur. Lungs congested.

Treatment.—Chloride of sodium or lime in dilute solution, and ordinary treatment for irritant poisoning.

Fatal Period (Shortest).—Fifteen minutes.

XVIII

BARIUM SALTS

Chloride of Barium occurs crystallized in irregular plates, like magnesium sulphate, soluble in water and bitter in taste. **Carbonate of Barium** is found in shops as a fine powder, tasteless and colourless, insoluble in water, but effervescing with dilute acids, and readily decomposed by the free acids of the stomach. **Nitrate of Barium** occurs in octahedral crystals, soluble in water.

Method of Extraction from the Stomach.—Dialysis as for other soluble poisons.

Tests.—Precipitated from its solutions by potassium carbonate or sulphuric acid. Burnt on platinum-foil, it gives a green colour to the flame.

Symptoms.—Besides those of irritants generally, violent cramps and convulsions, headache, debility, dimness of sight, double vision, noises in the ears, and beating at the heart. The salts of barium are also cardiac poisons.

Post-Mortem Appearances.—As of irritants generally. Stomach may be perforated.

Treatment.—Wash out stomach with a solution of sodium or magnesium sulphate, or of alum, and give stimulants by the mouth and hypodermically.

XIX

IODINE—IODIDE OF POTASSIUM

Iodine occurs in scales of a dark bluish-black colour. It strikes blue with solution of starch, and stains the skin and intestines yellowish-brown. Liquid preparations, as the liniment or tincture, may be taken accidentally or suicidally.

Symptoms.—Acrid taste, tightness of throat, epigastric pain, and then symptoms of irritant poisons generally. Chronic poisoning (iodism) is characterized by coryza, salivation, and lachrymation, frontal headache, loss of appetite, marked mental depression, acne of the face and chest, and a petechial eruption on the limbs.

Post-Mortem Appearances.—Those of irritant poisoning with corrosion, and staining of a dark brown or yellow colour.

Treatment.—Stomach-pump and emetics, carbonate of sodium, amylaceous fluids, gruel, arrowroot, starch, etc.

Analysis of Organic Mixture containing Iodine.—Add bisulphide of carbon, and shake. The iodine may be obtained on evaporation as a sublimate. It will be recognized by the blue colour which it gives with starch.

Iodide of Potassium.—Colourless, generally opaque, cubic crystals, soluble in less than their weight of cold water.

Symptoms.—Not an active poison, but even small doses sometimes produce the effects of a common cold, including those symptoms already mentioned as occurring with iodine.

Analysis.—Iodide of potassium in solution gives a bright yellow precipitate with lead salts; a bright scarlet with corrosive sublimate; and a blue colour with sulphuric or nitric acid and starch.

XX

PHOSPHORUS

Phosphorus is usually found in small, waxy-looking cylinders, which are kept in water to prevent oxidation. It may also occur as the amorphous non-poisonous variety, a red opaque infusible substance, insoluble in carbon disulphide. Ordinary phosphorus is soluble in oil, alcohol, ether, chloroform, and carbon disulphide; insoluble in water. It is much used in rat poisons, made into a paste with flour, sugar, fat, and Prussian blue. Yellow phosphorus is not allowed to be used in the manufacture of lucifer matches, and the importation of such is prohibited. In 'safety' matches the amorphous phosphorus is on the box.

Symptoms.—At first those of an irritant poison, but days may elapse before any characteristic symptoms appear, and these may be mistaken for those of acute yellow atrophy of the liver. The earliest signs are a garlicky taste in the mouth and pain in the throat and stomach. Vomited matter luminous in the dark, bile-stained or bloody, with garlic-like odour. Great prostration, diarrhœa, with bloody stools. Harsh, dry, yellow skin, purpuric spots with ecchymoses under the skin and mucous membranes, retention or suppression of urine, delirium, convulsions, coma, and death. Usually there are remissions for two to three days, then jaundice comes on, with enlargement of the liver; hæmorrhages from the mucous surfaces and under the skin; later, coma and convulsions. In chronic cases there is fatty degeneration of most of the organs and tissues of the body. The inhalation of the fumes of phosphorus, as in making vermin-killers, etc., gives rise to 'phossy-jaw.'

Post-Mortem Appearances.—Softening of the stomach, hæmorrhagic spots on all organs and under the skin, fatty degeneration of liver, kidneys, and heart, blood-stained urine, phosphorescent contents of alimentary canal.

Treatment.—Early use of stomach-pump and emetics, followed by the administration of permanganate of potassium or peroxide of hydrogen to oxidize the phosphorus. Oil should not be given. Sulphate and carbonate of magnesium, mucilaginous drinks. Sulphate of copper is a valuable antidote, both as an emetic and as forming an insoluble compound with phosphorus.

Fatal Dose.—One grain and a half.

Fatal Period.—Four hours; more commonly two to four days.

Detection of Phosphorus in Organic Mixtures.—Mitscherlich's method is the best. Introduce the suspected material into a retort. Acidulate with sulphuric acid to fix any ammonia present. Distil in the dark, through a glass tube kept cool by a stream of water. As the vapour passes over and condenses, a flash of light is perceived, which is the test.

XXI

ARSENIC AND ITS PREPARATIONS

Arsenic is the most important of all the metallic poisons. It is much used in medicine and the arts. It occurs as metallic arsenic, which is of a steel-grey colour, brittle, and gives off a garlic-like odour when heated; as arsenious acid; in the form of two sulphides—the red sulphide, or realgar, and the yellow sulphide, or orpiment; and as arsenite of copper, or Scheele's green. It also exists as an impurity in the ores of several metals—iron, copper, silver, tin, zinc, nickel, and cobalt. Sulphuric acid is frequently impregnated with arsenic from the iron pyrites used in preparing the acid. It is a constituent of many rat pastes, vermin or weed killers, complexion powders, sheep dips, etc.

Arsenious Acid (White Arsenic, Trioxide of Arsenic).—Colourless, odourless, and almost tasteless. It occurs in commerce as a white powder or in a solid cake, which is at first translucent, but afterwards becomes opaque. Slightly soluble in cold water; 1 ounce of water dissolves about 1/2 grain of arsenic. Fowler's solution is the best-known medicinal preparation of arsenic, and contains 1 grain of arsenious anhydride in 110 minims.

Symptoms.—Commence in from half to one hour. Faintness, nausea, incessant vomiting, epigastric pain, headache, diarrhœa, tightness and heat of throat and fauces, thirst, catching in the breath, restlessness, debility, cramp in the legs, and convulsive twitchings. The skin becomes cold and clammy. In some cases the symptoms are those of collapse, with but little pain, vomiting, or diarrhœa. In others the patient falls into a deep sleep, while in the fourth class the symptoms resemble closely those of English cholera. The vomited matters are often blue from indigo, or black from soot, or greenish from bile, mixed with the poison. Should the patient survive some days, no trace of arsenic may be found in the body, as the poison is rapidly eliminated by the kidneys. In all suspected cases the urine should be examined.

The symptoms of chronic poisoning by arsenic are loss of appetite, silvery tongue, thirst, nausea, colicky pains, diarrhœa, headache, languor, sleeplessness, cutaneous eruptions, soreness of the edges of the eyelids, emaciation, falling out of the hair, cough, hæmoptysis, anæmia, great tenderness on pressure over muscles of legs and arms, due to peripheral neuritis, and convulsions.

Pigmentation is common; the face becomes dusky red, the rest of the body a dark brown shade. This darkening is most marked in situations normally pigmented and in parts exposed to pressure of the clothes, such as the neck, axilla, and inner aspect of the arms, the extensor aspects being less marked than the flexor. The pigmentation resembles the bronzing of Addison's disease, but there are no patches on the mucous membranes, and the normal rosy tint of the lips is not altered. The skin over the feet may show marked hyperkeratosis.

The nervous system is notably affected. The sensory symptoms appear first: numbness and tingling of the hands and feet, pain in the soles of the feet on walking, pain on moving the joints, and erythromelalgia. Then come the motor symptoms, with drop-wrist and drop-foot. The patient suffers severely from neuritis, and there may be early loss of patellar reflex. The nervous symptoms come on later than the cutaneous manifestations.

Post-Mortem Appearances.—Signs of acute inflammation of stomach, duodenum, small intestines, colon, and rectum. Stomach may contain dark grumous fluid, and its mucous coat presents the appearance of crimson velvet. Ulceration is rare, and cases of perforation still less common, the patient dying before it occurs. If life has been preserved for some days, there is extensive fatty degeneration of the organs. There may be entire absence of post-mortem signs. Putrefaction of the body is retarded by arsenic.

Treatment.—The stomach-pump, emetics, then milk, milk and eggs, oil and lime-water. Inflammatory symptoms, collapse, coma, etc., must be treated on ordinary principles. As an antidote, the best when the poison is in solution is the hydrated sesquioxide of iron, formed by precipitating tinctura ferri perchloridi with excess of ammonia, or carbonate of soda. This is filtered off through muslin and given in tablespoonful doses. It forms ferric arsenate, which is sparingly soluble. Colloidal iron hydroxide may be used instead. Dialyzed iron in large quantities is efficacious.

Fatal Dose (Smallest).—Two grains. Exceptionally, recovery from very large doses if rejected by vomiting.

Fatal Period (Shortest).—Twenty minutes. Exceptionally, death as late as the sixteenth day. The effects of arsenic are modified by tolerance, some persons being able to take considerable quantities. The peasants of Styria are in the habit of eating it.

Method of Extraction from the Stomach.—The coats of the stomach should be examined with a lens for any white particles. These, if present, may be collected, mixed with a little charcoal in a test-tube, and heated. If arsenic is present, a metallic ring will be formed in the cooler parts of the tube. If this ring be also heated, octahedral crystals of arsenic will be deposited farther up the tube, and are easily recognized by the microscope. The contents of the stomach, or the solid organs minced up, should be boiled with pure hydrochloric acid and water, then filtered. The filtrate can then be subjected to Marsh's or Reinsch's process.

Tests.—In solution, arsenic may be detected by the liquid tests. (1) Ammonio-nitrate of silver gives a yellow precipitate (arsenite of silver). (2) Ammonio-sulphate of copper gives a green precipitate (Scheele's green). (3) Sulphuretted hydrogen water gives a yellow precipitate.

Marsh's Process.—Put pure distilled water into a Marsh's apparatus with metallic zinc and sulphuric acid. Hydrogen is set free, and should be tested by lighting the issuing gas and depressing over it a piece of white porcelain. If no mark appears, the reagents are pure, and the suspected liquid may now be added. The hydrogen decomposes arsenious acid, and forms arseniuretted hydrogen. The gas carried off by a fine tube is again ignited. A piece of glass or porcelain held to the flame will have, if arsenic be present, a deposit on it having the following characters: In the centre a deposit of metallic arsenic, round this a mixture of metallic arsenic and arsenious acid, and outside this another ring of arsenious acid in octahedral crystals. The deposit is dissolved by a solution of chloride of lime, turned yellow by sulphide of ammonium after evaporation; on the addition of strong nitric acid, evaporated and neutralized with ammonia and nitrate of silver added, a brick-red colour is produced—arseniate of silver.

Reinsch's Process.—Boil distilled water with one-sixth or one-eighth of hydrochloric acid, and introduce a slip of bright copper. If, after a quarter of an hour's boiling, there is no stain on the copper, add the suspected liquid. If arsenic be present, it will form an iron-grey deposit. If this foil be dried, cut up, put in a reduction-tube, and heated, crystals of arsenious trioxide will be deposited on the cold part of the tube.

These tests are difficult to apply, but as arsenic is a ubiquitous poison, and as there are many sources of fallacy, it would be well, when possible, to obtain the services of an expert.

Biological Test.—Put the substance to be tested into a flask with some small pieces of bread, sterilize for half an hour at 120° C. When cold, inoculate with a culture of Penicillium brevicaule, and keep at a temperature of 37° C. If arsenic is present, a garlic-like odour is noticed in twenty four hours, due to arseniuretted hydrogen or an organic combination of arsenic. This test is delicate, and will detect 1/1000 of a milligramme, but it is not quantitative.

Other Preparations of Arsenic.—These are arsenite of potash (Fowler's solution), cacodylate of sodium, and arsenite of copper (Scheele's green), the last frequently used for colouring dresses and wall-papers. Persons using these preparations may suffer from catarrhal symptoms, rashes on the neck, ears, and face, thirst, nausea, pain in stomach, vomiting, headache, perhaps peripheral neuritis and loss of patellar reflex. The cacodylates, although formerly employed in the treatment of phthisis, should be used with the utmost caution. The arsenites give the reactions of arsenious acid.

Arsenic is eliminated not only by the kidneys and bowels, but by the skin, and in women by the menses. It may be detected in the sweat, the saliva, the bronchial secretion, and, during lactation, in the milk.

The sale of arsenic and its preparations to the public is properly hedged round with restrictions of all kinds. It is included in Part I. of the Poisons and Pharmacy Act (8 Edward VII., c. 55). No arsenic may be sold to a person under age, nor may it be sold unless mixed with soot or indigo in the proportion of 1 ounce of soot or 1/2 ounce of indigo at the least to every pound of arsenic.

Arseniuretted Hydrogen (arsine, AsH_3) is an extremely poisonous gas, and is evolved in various chemical and manufacturing processes. When damp, Ferro-silicon evolves AsH_3 and PH_3, both very lethal gases. Ferrochrome is used in making steel, and it also evolves PH3, and in such extreme dilution as 0.02 per cent. may cause death.

XXII

ANTIMONY AND ITS PREPARATIONS

Tartar Emetic (tartarized antimony, potassio-tartrate of antimony) occurs as a white powder, or in yellowish-white efflorescent crystals. Vinum antimoniale contains 2 grains to a fluid ounce of the wine.

Symptoms.—Metallic taste, rapidly followed by nausea, incessant vomiting, burning heat and pain in stomach, purging. Dysphagia, sense of constriction in throat, intense thirst, cramps, faintness, profound depression; in fatal cases, giddiness and tetanic spasms. In chronic poisoning, nausea, vomiting and purging, weak pulse, loss of appetite, debility, cold sweats, great prostration, progressive emaciation. The symptoms in chronic poisoning may simulate gastritis or enteritis. Externally applied, it produces an eruption not unlike that of smallpox.

Post-Mortem Appearances.—Inflammation, softening, and an aphthous condition of the throat, gullet, and stomach, the last reddened in patches. In chronic poisoning, inflammation also of cæcum and colon. Brain and lungs may be congested. Decomposition is hindered for long.

Treatment.—Promote vomiting by warm greasy water, or the stomach-tube may be used. Cinchona bark or any preparation containing tannin, as tea, decoction of oak bark, etc. Morphine to allay pain.

Fatal Dose.—In an adult 2 grains (same as arsenic).

Fatal Period.—Death follows in eight to twelve hours, from exhaustion.

Method of Extraction from the Stomach.—The contents of the stomach or its coats should be finely cut up and boiled in water, acidulated with tartaric acid and subjected to dialysis, or strained and filtered. Pass hydrogen sulphide through the filtered or dialyzed fluid until a precipitate ceases to fall; collect the sulphide thus formed, wash and dry it. Boil the orange-coloured sulphide in a little hydrochloric acid. If the solution be now added to a large bulk of water, the white oxychloride is precipitated, which is soluble in tartaric acid and precipitated orange yellow with hydrogen sulphide. The chloride of bismuth is also precipitated white, but the precipitate is not soluble in tartaric acid, and the precipitate with hydrogen sulphide is black.

Tests.—Soluble in water, but not in alcohol.

Heated in substance, it crepitates and chars; and if heat be increased, the metal is deposited. Treated with sulphuretted hydrogen, a characteristic orange-red sulphide is formed.

A drop of the solution evaporated leaves crystals, either tetrahedric, or cubes with edges bevelled off. Sulphuretted hydrogen passed through gives the orange-red precipitate above named. Dilute nitric acid gives a white precipitate, soluble in excess, and also in tartaric acid. Marsh's and Reinsch's processes are applicable for the detection of antimony, but Reinsch's is the better. Reinsch's process gives a violet deposit instead of the black, lustrous one of arsenic.

Chloride of Antimony (Butter of Antimony).—A light yellow or dark red corrosive liquid.

Symptoms.—Violet corrosion and irritation of the alimentary canal, with the addition of narcotic symptoms. After death the mucous membrane of the entire canal is charred, softened, and abraded.

Treatment.—As for tartar emetic; magnesia in milk.

XXIII

MERCURY AND ITS PREPARATIONS

The most important salt of mercury, toxicologically, is corrosive sublimate. Other poisonous preparations are red precipitate, white precipitate, mercuric nitrate, the cyanide and potassio-mercuric iodide. Calomel has very little toxic action. Metallic mercury is not poisonous, but its vapour is.

Corrosive Sublimate (perchloride of mercury) is in heavy colourless masses of prismatic crystals, possessing an acrid, metallic taste. It is soluble in sixteen parts of cold and two of boiling water. Soluble in alcohol and ether, the latter also separating it from its solution in water.

Symptoms come on rapidly. Acrid, metallic taste, constriction and burning in throat and stomach, nausea, vomiting of stringy mucus tinged with blood, tenesmus, purging. Feeble, quick, and irregular pulse, dysuria with scanty, albuminous or bloody urine or total suppression. Cramp, twitches and convulsions of limbs, occasionally paralysis. In poisoning from the medicinal use of mercury, there may be salivation, a coppery taste in the mouth, peculiar fœtor of breath, tenderness and swelling of mouth, inflammation, swelling and ulceration of gums (cancrum oris), a blue line on the gums, and the loosening of teeth. Mercury is less quickly eliminated from the body than arsenic. In chronic cases 'mercurialism,' 'hydrargyrism,' 'ptyalism,' or 'salivation,' including most of the symptoms enumerated above. May get eczema mercuriale and periostitis. Profound anæmia often a prominent symptom; neuritis not uncommon. If fumes of mercury inhaled, mercurial tremors develop.

Post-Mortem Appearances.—Corrosion, softening, and sloughing ulceration of stomach and intestines. The mucous membrane of the œsophagus and stomach is often of a bluish-grey colour. The large intestine and rectum are often ulcerated and gangrenous. Inflamed condition of urinary organs, with contraction of the bladder.

Treatment.—Encourage or produce vomiting. Albumin, as white of egg, gluten, or wheat flour, is the best antidote. Demulcent drinks, milk, and ice. Stomach-tube to be used with care, owing to softened state of gullet and stomach.

Fatal Dose.—Three grains in a child.

Fatal Period.—Half an hour the shortest.

Method of Extraction from the Stomach.—A trial test may be made of the contents of the stomach with copper-foil. If mercury is found, the contents of the stomach may be dialyzed, the resulting clear fluid concentrated and shaken with ether, which has the power of taking corrosive sublimate up, and thus separating it from arsenic and other metallic poisons. The ether allowed to evaporate will leave the corrosive sublimate in white silky-looking prisms. Suppose no mercury is found in the dialyzed fluid, owing to the fact that corrosive sublimate enters into insoluble compounds with albumin, fibrin, mucous membrane, gluten, tannic acid, etc., we must dry the insoluble matter, and heat it with nitro-hydrochloric acid until all organic matter is destroyed and excess of nitric acid expelled. The residue dissolved in water, filtered, and tested with copper-foil, etc.

Tests.—The following table gives the action of corrosive sublimate with reagents:

1. With iodide of potassium	Bright scarlet colour.
2. With potash solution	Bright yellow colour.
3. With hydrochloric acid and sulphuretted hydrogen	First a yellowish and then a black colour.
4. Heated in a reduction-tube	Melts, boils, is volatilized, and forms a white crystalline sublimate.
5. With ether	Freely soluble; the ethereal solution, when allowed to evaporate spontaneously, deposits the salt in white prismatic crystals.

*6. Heated with carbonate of
sodium in a reduction-tube
Globules of metallic mercury
are produced.*

A very simple process for detecting corrosive sublimate is to put a drop of the suspected solution on a sovereign and touch the gold through the solution with a key, when metallic mercury will be deposited on the gold.

XXIV

LEAD AND ITS PREPARATIONS

Acetate of Lead (Sugar of Lead).—A glistening white powder or crystalline mass. Soluble in water, with a sweetish taste. It is practically the only lead salt which gives rise to acute symptoms, and only when taken in large doses.

Symptoms.—Metallic taste, dryness in throat, intense thirst, vomiting, colicky pains, cramps, cold sweat, constipation and scanty urine, severe headache, convulsions.

Chronic lead-poisoning is liable to occur in those who handle lead in any form—white-lead workers, paint manufacturers, plumbers, pottery workers, etc.

In chronic lead-poisoning the most prominent symptoms are a blue line on the gums, anæmia, emaciation, pallor, quick pulse, persistent constipation, colic, cramps in limbs, and paralysis of the extensor muscles, causing 'dropped hand.' May get saturnine encephalopathies, of which intense headache, optic neuritis, and epileptiform convulsions, are the most common. Albumin in urine, tendency to gout, and in women to abortion.

Post-Mortem Appearances.—Inflamed mucous membrane of stomach and intestines, with layers of white or whitish-yellow mucus, impregnated with the salt of lead.

Treatment.—Sulphate of sodium or magnesium, or a mixture of dilute sulphuric acid, spirits of chloroform, and peppermint-water. Milk, or milk and eggs. As a prophylactic among workers in lead, a drink containing sulphuric acid flavoured with treacle should be given. Lavatory accommodation should be provided, and scrupulous cleanliness should also be enjoined in the workshops. The dry grinding of lead salts should be prohibited. The ionization method of Sir Thomas Oliver is most useful both as regards cure and also prevention of chronic poisoning by lead.

Fatal Dose and Fatal Period.—Uncertain.

Method of Extraction from the Stomach.—Dry the contents of the stomach or portions of the liver, etc., and incinerate in a porcelain crucible. Treat the ash with nitric acid, dry, and dissolve in water. The solution of nitrate of lead may now have the proper tests applied.

Tests.—Sulphuretted hydrogen gives a black precipitate; liquor potassæ, white precipitate; sulphuric acid, white precipitate, insoluble in nitric acid; iodide of potassium, a bright yellow precipitate. A delicate test for lead in water is to stir the water, concentrated or not, with a glass rod dipped in ammonium sulphide: a brown coloration is produced. One-tenth of a grain of lead in a gallon of water may be detected.

Chronic lead-poisoning is an 'industrial disease,' and, being an occupation risk, its victims are entitled to compensation at the hands of their employers. In case of death, compensation has been awarded even when at the autopsy the patient has been found to have suffered from acute tuberculosis of the lungs. The responsibility of apportioning the monetary value of disablement resulting from the action of the lead rests with a judge or jury, who are guided by the expert medical evidence available.

Diachylon, or lead-plaster, is largely used as an abortifacient.

XXV

COPPER AND ITS PREPARATIONS

Poisoning with copper salts is rare. The most important are the sulphate, subacetate, and arsenite.

Sulphate of Copper (bluestone, blue vitriol) in half-ounce doses is a powerful irritant. Has been given to procure abortion.

Subacetate of Copper (verdegris) occurs in masses, or as a greenish powder. Powerful, astringent, metallic taste. Half-ounce doses have proved fatal.

Symptoms.—Epigastric pain, vomiting of bluish or greenish matter, diarrhœa. Dyspnœa, depression, cold extremities, headache, purple line round the gums. Jaundice is common. A chronic form of poisoning may occur, with symptoms closely resembling those of lead.

Post-Mortem Appearances.—Inflammation of stomach and intestines, which are bluish or green in colour.

Treatment.—Encourage vomiting. Give albumin or very dilute solution of ferrocyanide of potassium.

Method of Extraction from the Stomach.—Boil the contents of the stomach in water, filter, pass hydrogen sulphide, filter, collect precipitate and boil in nitric acid, filter, dilute filtrate with water and apply tests. In the case of the solid organs, dry, incinerate, digest ash in hydrochloric acid, evaporate nearly to dryness, dilute with water, and test.

Tests.—Polished steel put into a solution containing a copper salt receives a coating of metallic copper. Ammonia gives a whitish-blue precipitate, soluble in excess. Ferrocyanide of potassium gives a rich red-brown precipitate. Sulphuretted hydrogen gives a deep brown precipitate.

XXVI

ZINC, SILVER, BISMUTH, AND CHROMIUM

The salts of zinc requiring notice are the sulphate and chloride.

Sulphate of Zinc has been taken in mistake for Epsom salts. In large doses it causes dryness of throat, thirst, vomiting, purging, and abdominal pain.

Post-Mortem Appearances.—Those of inflammation of digestive tract.

Treatment.—Tea, decoction of oak-bark, carbonate of potassium or sodium as antidote.

Chloride of Zinc.—A solution containing this substance (230 grains to the ounce) constitutes 'Burnett's disinfecting fluid.' It is a corrosive poison.

The symptoms are burning sensation in the mouth, throat, stomach, and abdomen, followed by vomiting, diarrhœa, with tenesmus and distension of the abdomen. The vomited matter contains shreds of mucous membrane with blood. There is profound collapse, cold surface, clammy sweats, weak pulse, with great prostration. The treatment is to wash out the stomach with large and weak solutions of carbonate of sodium. Mucilaginous drinks may be given, and hypodermic injections of morphine are useful to allay the pain.

Method of Extraction from the Stomach.—Dry and incinerate the tissues in a porcelain crucible, digest ash in water, apply tests.

Tests.—Ammonia, a white precipitate soluble in excess, reprecipitated by sulphuretted hydrogen; ferrocyanide of potassium, a white precipitate; sulphuretted hydrogen, a white precipitate in pure and neutral solutions. Nitrate of baryta will show the presence of sulphuric acid, and nitrate of silver of hydrochloric acid.

Silver.—Nitrate of silver is a powerful irritant.

Tests.—Black precipitate with sulphuretted hydrogen; white with hydrochloric acid.

Treatment.—Common salt.

Chronic nitrate of silver poisoning is characterized by argyria. The gums show a blue line, which is darker than that produced by lead, and the skin presents a greyish hue, which is permanent.

Bismuth.—The bismuth salts are not poisonous, but may contain arsenic as an impurity, although this is far less common than it was some years ago.

Chromic Acid, Chromate, Bichromate of Potassium.—These act as corrosives when solid or in concentrated liquid forms. In dilute solutions they act as irritants. Used as dyes; have proved fatal more than once. Those engaged in their manufacture suffer from unhealthy ulcers on the nasal septum and hands. The former may to some extent be prevented by taking snuff. Lead chromate (chrome yellow) is a powerful irritant poison. Two drachms of the bichromate caused death in four hours.

Tests.—Yellow precipitate with salts of lead, deep red with those of silver.

Treatment.—Emetics, magnesia, and diluents. Washing out of the stomach with weak solution of nitrate of silver.

XXVII

GASEOUS POISONS

Carbon Dioxide.—Carbon dioxide is a product of combustion and respiration, and is generated in many ways during fermentation. It is a constituent of choke damp due to explosions in coal-mines, and is given off from lime-kilns, brick-kilns, and cement-works. It is often met with in dangerous quantities in wells and in brewers' vats. From 10 to 15 per cent. in the atmosphere would prove fatal, but even 2 per cent. inhaled for long would produce serious symptoms. The proportion usually present in air is 0.04 per cent.

Symptoms.—Inhalation of the pure gas causes spasm of the glottis, insensibility, and death from asphyxia, at once; diluted, causes sense of weight in forehead and back of head, giddiness, vomiting, somnolence, loss of muscular power. Insensibility, stertorous breathing, lividity of face and body, and death from asphyxia. Convulsions occasionally.

Post-Mortem Appearances.—Face swollen and livid, or calm and pale; lividity is most marked in eyelids, lips, ears, etc.; limbs usually flaccid, abdomen distended; right side of heart, lungs, and large veins, gorged with dark-coloured blood. Brain and membranes congested.

Treatment.—Pure air, cold affusion, stimulants, artificial respiration, galvanism, inhalation of oxygen, venesection, transfusion.

Carbonic Oxide.—This is one of the most poisonous of gases. It is evolved in the process of burning charcoal and coke in stoves or furnaces. Water-gas, obtained by passing steam over heated coke, contains 40 per cent. of the substance, the remainder being chiefly hydrogen. It forms the chief part of the deadly 'choke damp' after an explosion in a mine. Two per cent. in the atmosphere is immediately fatal.

Symptoms.—When in large amount, insensibility comes on at once; when in very small amounts, headache, giddiness, noises in the ears, nausea, and vomiting, with prostration, insensibility, and coma. There may be convulsions. Even in cases which recover, permanent impairment of the brain may result.

Post-Mortem Appearances.—The blood is bright red in colour, due to the interaction of carbonic oxide with hæmoglobin. A rosy hue of the skin-surface and viscera is often noticed. Bright red patches of colour are found over the surface of the body. The spectrum of the blood is characteristic.

Treatment.—Ammonia to the nostrils, inhalation of oxygen, cold douche in moderation, artificial respiration, transfusion of blood.

Coal Gas.—Coal gas contains light carburetted hydrogen or marsh gas, olefiant gas, ammonia, sulphuretted hydrogen, carbonic acid, carbonic oxide, free hydrogen, and nitrogen. Coal gas has an offensive odour, burns with a yellowish-white flame, yielding water and carbonic acid. Cases of poisoning often due to escape of gas into the room.

Symptoms.—Headache and giddiness, foaming at mouth, vomiting, convulsions, tetanic spasms, stertorous breathing, dilated pupil. The breath smells of gas; there is profound stupor; the patient, if alive, exhales gas from the lungs when removed into a fresh room or into the air. Smell of gas in the room and in patient's breath.

Post-Mortem Appearances.—Pallor of skin and internal tissues; florid colour of neck, back, and muscles, if much CO present in the coal gas; fluid florid blood; infiltration of lungs.

Treatment.—Fresh air, artificial respiration, cold affusion, diffusible stimulants; inhalation of oxygen freely.

Sulphuretted Hydrogen is characterized by its odour, like that of rotten eggs. It is extremely poisonous.

Symptoms.—Giddiness, pain and oppression in stomach, nausea, loss of power; delirium, tetanus, and convulsions.

Post-Mortem Appearances.—Fluid and black blood (sulph-hæmoglobin), smell of H2S on opening the body; loss of contractility of muscles, rapid putrefaction.

Treatment.—Fresh air, stimulants, inhalation of chlorine.

Tests.—Acetate of lead throws down a brown or black precipitate according to the quantity of the gas.

Sewer Gas.—Cesspool emanations usually consist of a mixture of sulphuretted hydrogen, sulphide of ammonium, and nitrogen; but sometimes it is only deoxidized air with an excess of carbonic acid gas.

Symptoms.—If poison concentrated, death may ensue at once; if gas diluted, or exposure only short, insensibility, lividity, hurried respiration, weak pulse, dilated pupils, elevation of temperature to 104°, tonic convulsions not unlike those of tetanus.

Treatment.—Fresh air, oxygen, with artificial respiration. Stimulants, hypodermic of strychnine, and alternate hot and cold douche.

Irritant Gases are—(1) Nitrous acid gas; (2) sulphurous acid gas; (3) hydrochloric acid gas; (4) chlorine; (5) bromine; (6) ammonia. They have the common property of causing irritation and inflammation of the eyes, throat, and air-passages, and may cause spasm of the glottis, bronchitis, and pneumonia.

Sulphurous Acid Gas.—One of the products of combustion of common coal.

Hydrochloric Acid Gas.—Irrespirable when concentrated, and very irritating when diluted. Very destructive to vegetable life.

Chlorine.—Used in bleaching, and as a disinfectant. Greenish-yellow colour, suffocating odour. In poisoning, inhalation of sulphuretted hydrogen gives relief.

XXVIII

VEGETABLE IRRITANTS

The chief vegetable purgatives are aloes, colocynth, gamboge, jalap, scammony, seeds of castor-oil plant, croton-oil, elaterium, the hellebores, and colchicum. All these have, either alone or combined, proved fatal. The active principle in aloes is aloin; of jalap, jalapin; of white hellebore, veratria; and of colchicum, colchicin. Morrison's pills contain aloes and colocynth; aloes is also the chief ingredient in Holloway's pills.

Symptoms.—Vomiting, purging, tenesmus, etc., followed by cold sweats, collapse, or convulsions.

Post-Mortem Appearances.—Inflammation of alimentary canal; ulceration, softening, and submucous effusion of dark blood.

Treatment.—Diluents, opium, stimulants, abdominal fomentations, etc.

Certain of these irritant poisons exert a marked influence on the central nervous system, as the following:

Laburnum (Cytisis Laburnum).—All parts of the plant are poisonous; the seeds, which are contained in pods, are often eaten by children. Contains the alkaloid cytisine, which is also contained in arnica. It has a bitter taste, and is powerfully toxic. Symptoms are purging, vomiting, restlessness, followed by drowsiness, insensibility, and convulsive twitchings. Death due to respiratory paralysis. Most of the cases are in children. Treatment consists of stomach-pump or emetics, stimulants freely, artificial respiration, warmth and friction to the surface of the body.

Yew (Taxus baccata) contains the alkaloid taxine. The symptoms are convulsions, insensibility, coma, dilated pupils, pallor, laboured breathing, collapse. Death may occur suddenly. Treatment as above. Post-mortem appearances not characteristic, but fragments of leaves or berries may be found in the stomach and intestines.

Arum (Arum Maculatum).—This plant, commonly known as 'lords and ladies,' is common in the woods, and the berries may be eaten by children. It gives rise to symptoms of irritant poisoning, vomiting, purging, dilated pupils, convulsions, followed by insensibility, coma, and death.

Many plants have an intensely irritating action on the skin, and when absorbed act as active poisons.

Rhus toxicodendron is the poison oak or poison ivy. Poisoning by this plant is rare in England, though not uncommon in the United States. Mere contact with the leaves or branches will in many people set up an acute dermatitis, with much œdema and hyperæmia of the skin. The inflammation spreads rapidly, and there is formation of blebs with much itching. There is often great constitutional disturbance, nausea, vomiting, diarrhœa, and pains in the abdomen. The effects may last a week, and the skin may desquamate.

Primula obconica is another plant which, when handled, gives rise to an acute dermatitis of an erysipelatous character. The face swells, and large blisters form on the cheeks and chin.

XXIX

OPIUM AND MORPHINE

Opium.—The inspissated juice of the unripe capsules of the Papaver somniferum. As a poison it is generally taken in the form of the tincture (laudanum), which contains 1 grain opium in 15 minims. Opium is found in almost all so-called 'soothing syrups' for children, and in Godfrey's cordial, Dalby's carminative, and Collis Browne's chlorodyne. Laudanum contains 1 per cent. morphine, and it, along with all other preparations (e.g., paregoric) which contain 1 or more per cent. morphine, are included in Part I. of the Schedule of Poisons, and come under the Dangerous Drugs Regulations.

The most important active principles of opium are the alkaloids morphine and codeine.

Symptoms usually commence in from twenty to thirty minutes: Giddiness, drowsiness and stupor, followed by insensibility. Patient seems asleep; may be roused by loud noise, but quickly relapses. Breathing slow and stertorous, pulse weak, countenance livid. As coma increases, pulse becomes slower and fuller. The pupils are contracted, even to a pin's point; they are insensible to the action of light. In deep, natural sleep the eyes are turned upwards and the pupils contracted. Bowels confined, skin cold and livid or bathed in sweat. Temperature subnormal. Nausea and vomiting are sometimes present. Remissions are not infrequent, the patient appearing about to recover and then relapsing. Hæmorrhage into the pons may give rise to contracted pupils. Young children and infants are specially susceptible to the poison.

Diagnosis is not always easy, and one has to differentiate poisoning from cerebral apoplexy. In the latter one can seldom rouse the patient, the pupils are often unequal, and hemiplegia is present. In compression of the brain, fracture of the skull may be present, subconjunctival hæmorrhages may be seen, the pupils are unequal and dilated, and the paralysis increases. In uræmic or diabetic coma the urine must be examined.

The habitual use of opium is not uncommon, and opium-eaters are able to take enormous quantities of the drug. The opium-eater may be known by his attenuated body, withered yellow countenance, stooping posture, and glassy, sunken eyes.

Post-Mortem Appearances.—Not characteristic. Turgescence of cerebral vessels. There may be effusion under arachnoid, into ventricles, at base of the brain, and around the cord. Rarely extravasation of blood. Stomach and intestines usually healthy. Lungs gorged, skin livid.

Fatal Period.—Usually nine to twelve hours; but in many cases, if life is prolonged for eight hours, recovery takes place.

Fatal Dose.—Four grains of opium is the smallest fatal dose in an adult, or one drachm of laudanum; children are proportionately much more susceptible to the action of opium than adults.

Treatment.—Stomach-tube, emetics, strong coffee or tea, ammonia to nostrils. Give 10 grains of permanganate of potassium in a pint of water acidulated with sulphuric acid, and repeat the dose every half hour. Belladonna by mouth, or atropine hypodermically. Patient must be kept roused by dashing cold water over him, flagellating with a wet towel, walking about, etc. In conditions of collapse, however, this treatment must not be continued, but everything should be done to preserve the strength. Treatment must be continued as long as life remains.

Method of Extraction from the Stomach.—Opium itself cannot be directly detected, but we test for morphine and meconic acid. These may be separated from organic mixtures thus: Boil the organic matter with distilled water, spirit, and acetic acid; filter, and to the fluid passed through add acetate of lead till precipitate ceases. Filter. Acetate of morphine passes through, and meconate of lead remains. The solution of acetate of morphine may be freed from excess of lead by hydrogen sulphide and filtered, excess of hydrogen sulphide driven off by heat, and tests applied. Put the meconate of lead with water into a beaker and pass hydrogen sulphide; sulphide of lead is formed, and meconic acid set free. Filter. Concentrate the solution of meconic acid, allow a portion to crystallize, and apply tests.

Tests.—Morphine and its acetate give an orange-red colour with nitric acid, becoming brighter on standing; decompose iodic acid, setting free iodine; with perchloride of iron, gives a rich indigo-blue; with bichromate of potassium, a green turning to brown. When the alkaloid is heated in a watchglass with a drop of strong sulphuric acid until the acid begins to fume, and is then allowed to get quite cold, a drop of nitric acid produces a brilliant red colour. The iodic acid test is very delicate, but requires great care, and may be used in the presence of organic matter.

Meconic acid gives a blood-red colour with perchloride of iron, not discharged by corrosive sublimate or chloride of gold. The similar colour produced by sulpho-cyanide of potassium and perchloride of iron is discharged by chloride of gold and corrosive sublimate.

Morphine Habit.—Individuals who have acquired this habit take the drug usually by hypodermic injection. The victim suffers from nausea and vomiting, and becomes so mentally debilitated that asylum treatment is required.

XXX

BELLADONNA, HYOSCYAMUS, AND STRAMONIUM

Belladonna.—The root, leaves, and berries, of the Atropa belladonna are poisonous from the presence of alkaloid atropine.

Symptoms.—Dryness of mouth and throat, intense thirst, dysphagia and dysphonia, quick pulse, noisy delirium and stupor. Strangury and hæmaturia, and redness of the skin, especially of the face, like that of scarlatina, have been noticed. Dilatation of the pupil occurs, whether the poison be taken internally or applied locally to the eye.

Post-Mortem Appearances.—Congestion of cerebral vessels, dilated pupils, red patches in alimentary canal.

Treatment.—Wash out the stomach freely; a hypodermic injection of apomorphine as an emetic, followed by hypodermic injections of pilocarpine or morphine. Tea, coffee, or tannin, to precipitate the alkaloid.

Tests.—Atropine may be recognized by its action on the pupil. The chloro-iodide of potassium and mercury precipitates it from very dilute solutions.

Hyoscyamus (Henbane).—Hyoscyamus niger.

Stramonium (Thorn-Apple).—Datura stramonium.

Symptoms.—Identical with those of belladonna and hyoscyamus, the post-mortem appearances and treatment being also the same.

Cannabis Indica (Indian Hemp).—When smoked, produces intoxication and mania. Hashish, used in the East as a narcotic, may cause persons to run 'amok' and commit murder.

XXXI

COCAINE

Cocaine.—Any dose above 1/2 grain applied to a mucous membrane or injected hypodermically may give rise to alarming symptoms. These are intense pallor, faintness, giddiness, dilatation of pupils, paroxysmal dyspnœa, rapid, intermittent, and weak pulse, nausea and vomiting, intense prostration verging on collapse, and convulsions. The patient may recover if allowed to remain in a recumbent position, but stimulants by mouth—e.g., ammonia—and the hypodermic injection of brandy or ether may be necessary, with the inhalation of nitrite of amyl.

For care in the prescribing of cocaine see under the 'Dangerous Drugs Act, 1920' (p. 82).

The **Cocaine Habit** consists in the self-administration of the drug hypodermically. It induces excitement, which is followed by prostration. In time melancholia or mania develops, with great irritation of the skin ('cocaine bugs').

XXXII

CAMPHOR

The liniment, oil, and spirit have been poisonous in large dose.

Symptoms.—Odour of breath, languor, giddiness, faintness, dimness of vision, difficulty of breathing, delirium, convulsions, with hot skin, flushed face, and dilated pupils.

Fatal Dose.—Thirty grains.

Cocculus Indicus.—The fruit of Anamirta cocculus. Contains a poisonous active principle, picrotoxin; used to adulterate beer, and by poachers to stupefy fish.

Symptoms.—Convulsions, followed by stupor and complete loss of voluntary power.

XXXIII

TETRACHLORETHANE, ETC.

Tetrachlorethane ('Cellon').—Acetylene tetrachloride; vapour has caused poisoning in aeroplane ('dope') and cinema film works.

Symptoms.—Gastric symptoms and marked jaundice. This may be followed in days or weeks by stupor, coma, death.

Post-Mortem.—Fatty degeneration of internal organs, chiefly liver.

Trinitrotoluene (T.N.T.).—An explosive solid which stains the skin an orange colour; may be absorbed through skin or be inhaled.

Symptoms.—Shortness of breath, headache, drowsiness. Later, skin irritation, gastritis, jaundice, blood degeneration.

Treatment.—Remove from work, rest in bed, diuretics, purgatives, alkalies.

XXXIV

ALCOHOL, ETHER, AND CHLOROFORM

Alcohol, ether, and chloroform, induce general anæsthesia, often preceded by delirious excitement, and followed by nausea and vomiting. When they cause death, it is by inducing a state like apoplexy or by paralyzing the heart.

Alcohol.—Absolute alcohol is ethyl hydroxide (C_2H_5OH) with not more than 1 per cent. by weight of water. Rectified spirit (spiritus rectificatus) contains 90 per cent. of alcohol. Methylated spirit consists of rectified spirit with 10 per cent. of wood spirit. Proof spirit contains a little over 49 per cent. of absolute alcohol; brandy or whisky, 53 per cent.; port wine, 20 to 25 per cent.; ales and stout, 4 to 6 per cent.

Symptoms.—Acute poisoning; confusion, giddiness, staggering gait, headache, passing into stupor, with subnormal temperature, and coma. Vomiting may occur and recovery ensue, otherwise collapse sets in. Pupils usually dilated.

Dipsomaniacs suffer from indigestion, vomiting and purging, jaundice, albuminuria, diabetes, cirrhosis of liver, degeneration of kidneys, congestion of brain, peripheral neuritis, alcoholic insanity, and various forms of paralysis. In the acute form delirium tremens is the most common manifestation.

Post-Mortem Appearances.—Deep red colour of lining membranes of stomach. Sometimes congestion of cerebral vessels and meninges. Lungs congested, blood fluid. Rigor mortis persistent.

Fatal Dose.—Death from 1/2 pint of gin and from two bottles of port, but recovery from larger quantities.

Fatal Period.—Average about twenty-four hours.

Treatment.—Stomach-tube, cold affusion, electricity, injection of a pint of hot coffee into the rectum. Give chloride of ammonium in 30 grain doses to prevent delirium; strychnine or digitalin hypodermically.

Method of Extraction from the Stomach.—Neutralize the contents of the stomach, if acid, with sodium carbonate; place them in a retort and carefully distil. Collect the distillate, mix with chloride of calcium or anhydrous sulphate of copper, and again distil. Agitate distillate with dry potassium carbonate, and draw off some of the supernatant fluid for testing.

Tests.—Odour. Dissolves camphor. With dilute sulphuric acid and bichromate of potassium turns green, and evolves aldehyde. Product of combustion makes lime-water white and turbid.

Methyl Alcohol: Wood Naphtha.—Used to produce intoxication by painters, furniture-polishers, etc.

Symptoms are those of alcoholic poisoning, but vomiting and delirium are more persistent. Total or partial blindness may follow as a sequel of optic atrophy. A fatal result not infrequently follows.

The following table gives the points of distinction between concussion of brain, alcoholic poisoning, and opium poisoning:

Concussion of Brain.	Alcohol.	Opium.
1. Marks of violence on head	1. No marks of violence, unless person has fallen. History will be of use.	1. As alcohol.
2. Stupor, sudden.	2. Excitement precedes sudden stupor.	2. Symptoms slow. Drowsiness, stupor, lethargy.
3. Face pale, cold; pupils sluggish, sometimes dilated.	3. Face flushed; pupils generally dilated.	3. Face pale; pupils contracted.
4. Remission rare. Patient recovers slowly.	4. Partial recovery may occur, followed by death.	4. Remission rare.
5. No odour of alcohol in breath.	5. Odour of alcohol in breath.	5. Odour of opium in breath.

Ether is a volatile liquid prepared from ethylic alcohol by interaction with sulphuric acid. It contains 92 per cent. of ethyl oxide $(C_2H_5)O$. It was formerly called 'sulphuric ether.' It is a colourless, inflammable liquid, having a strong and characteristic odour, specific gravity 0.735. **Purified ether** from which the ethylic alcohol has been removed by washing with distilled water, and most of the water by subsequent distillation in the presence of calcium chloride and lime. It is this preparation which is used for the production of general anæsthesia. It has a specific gravity of 0.722 to 0.720, and its vapour is very inflammable.

Symptoms.—When taken as a liquid, same as alcohol. When inhaled as vapour, causes slow, prolonged, and stertorous breathing; face becomes pale, lips bluish, surface of body cold. Pulse first quickens, then slows. Pupils dilated, eyes glassy and fixed, muscles become flabby and relaxed, profound anæsthesia. Then pulse sinks and coma ensues, sensation being entirely suspended. Nausea and vomiting not uncommon.

Post-Mortem Appearances.—Brain and lungs congested. Cavities of heart full of dark, liquid blood. Vessels at upper part of spinal cord congested.

Treatment.—Exposure to pure air, cold affusion, artificial respiration, galvanism.

Method of Extraction from the Contents of the Stomach.—Same as for alcohol. During distillation pass some of the vapour into concentrated solution of bichromate of potash, nitric and sulphuric acids, and note reaction as for alcohol.

Tests.—Vapour burns with smoky flame, depositing carbon. Sparingly soluble in water. With bichromate of potash and sulphuric acid same as alcohol.

Chloroform.—A colourless liquid, specific gravity 1.490 to 1.495, very volatile, giving off dense vapour. Sweet taste and pleasant odour.

Symptoms.—When swallowed, characteristic smell in breath, anxious countenance, burning pain in the throat, stomach, and region of the abdomen, staggering gait, coldness of the extremities, vomiting, insensibility, deepening into coma, with stertorous breathing, dilated pupils, and imperceptible pulse. When inhaled, much the same as ether, but produces insensibility and muscular relaxation more rapidly. It would be impossible to instantly render a person insensible by holding a pocket-handkerchief saturated with chloroform over the face. Statements such as this, which are often made in cases of robbery from the person and in cases of rape, are incredible.

Delayed Chloroform-Poisoning.—Death may take place in from four to seven days after chloroform has been administered, especially in the case of children. The internal organs are found to be fattily degenerated, and death is thought to be due to acetonuria.

Post-Mortem Appearances.—Cerebral and pulmonary congestion. Heart empty, or right side distended with dark blood.

Treatment.—Stomach-tube and free lavage; cold affusion; drawing forward tongue; artificial respiration; galvanism and suspension with head downward. Inhalation of nitrite of amyl; strychnine hypodermically.

Fatal Dose.—When swallowed, from 1 to 2 ounces.

Method of Extraction from the Stomach.—By distillation at 120° F. The vapour, as it passes along a glass tube, may be decomposed by heat into chlorine, hydrochloric acid, and carbon—the first shown by setting free iodine in iodide of starch, the second by reddening blue litmus-paper, and the last by its deposit.

Tests.—Taste, colour, weight; burns with a green flame; dissolves camphor, guttapercha, and caoutchouc.

Iodoform.—Poisoning may result from its use in surgery. It produces delirium, sleepiness, and coma. It may lead to mental weakness or optic neuritis.

XXXV

CHLORAL HYDRATE

It was formerly largely used as a hypnotic, and many fatal consequences ensued. It is prepared from alcohol and chlorine.

Symptoms.—Deep sleep, loss of muscular power, diminished or abolished reflex action and sensibility, followed by loss of consciousness and marked fall of temperature. Pulse may become quick, and face flushed or livid and bloated. Prolonged use of this drug may produce a peculiar eruption on the skin. Supposed to act in the blood by being decomposed into chloroform and sodium formate. Its effects are due chiefly to depression of the central nervous system, the medulla being the last part of the nervous system to be attacked.

Method of Extraction from the Stomach.—By distillation in strongly alkaline solutions, when it may be obtained as chloroform and tested as such.

Treatment.—Stomach-tube or emetic. Hypodermic injections of strychnine. Keep patient warm, and inject a pint of hot strong coffee into the rectum. Nitrite of amyl and artificial respiration.

Tests.—Heated with caustic potash, it yields chloroform and potassium formate. The chloroform is readily recognized by its odour, and, if the solution be concentrated, by separating as a heavy layer at the bottom of the test-tube.

XXXVI

PETROLEUM AND PARAFFIN-OIL

Cases of poisoning by petroleum and paraffin are common, and occur chiefly in children.

Petroleum is a natural product, and is a mixture of the higher saturated hydrocarbons. The crude petroleum is purified by distillation, and is then free from colour, but retains its peculiar penetrating odour. Different varieties are sold under the names of cymogene, gasolene, naphtha, petrol, and benzoline. Benzoline is highly inflammable, and is often called mineral naphtha, petroleum naphtha, and petroleum spirit. Benzoline is not the same as benzene or benzol, which is one of the products of the dry distillation of coal.

From its very general use as a fuel in motor-cars many accidents have happened from inhaling the vapour of petrol. It gives rise to coldness, shallow respiration, syncope, and insensibility, but seldom death.

Paraffin, also known as kerosene and mineral oil, is a mixture of saturated hydrocarbons obtained by the distillation of shale.

By the retailer the terms 'petroleum' and 'paraffin' oil are used indifferently, and each is sold for the other without prejudice.

Symptoms.—These substances are not very active poisons, and, as a rule, even children recover. The breath has the odour of paraffin, the face is pale and cyanotic, hot and dry, and there may be vomiting. Death may result from gastro-enteritis or from coma.

Fatal Dose.—In the case of an adult, 1/2 pint should not prove lethal, and patients have recovered after drinking a pint.

Treatment.—Emetics, purgatives, and stimulants.

XXXVII

ANTIPYRINE, ANTIFEBRIN, PHENACETIN, AND ANILINE

Many of the synthetical coal-tar products now so largely employed as analgesics are powerful toxic agents.

Phenazone, Antipyrine, or Analgesin, is a complex benzene derivative prepared from aniline, aceto-acetic ether, and methyl iodide. It is in colourless, inodorous, scaly crystals, which have a bitter taste. It is soluble in its own weight of water.

Tests.—Can be extracted from an alkaline solution of chloroform. The residue left on the evaporation of chloroform should be employed for testing. If heated with strong nitric acid and allowed to cool, a purple colour is produced. Ferric chloride gives a blood-red coloration, destroyed by the addition of mineral acids.

Treatment.—Stimulants freely, inhalation of oxygen, patient to be kept in the recumbent position.

Acetanilide, Antifebrin, Phenylacetamide (a constituent of 'Daisy' or 'headache' powders), is obtained by the interaction of acetic acid and aniline. It is in colourless, inodorous, lamellar crystals, which have a slight pungent taste. It is insoluble in water.

Tests.—May be extracted from acid solutions by ether or chloroform. If heated with solution of potassium hydroxide, odour of aniline is given off; if liquid, when it is warmed with a few drops of chloroform, a penetrating and unpleasant odour of isocyanide.

Treatment.—Emetics, stimulants, inhalation of ether, recumbent position.

Phenacetin, Phenacetinum, is produced by the interaction of glacial acetic acid and para-phenetidin. It is in white, tasteless, inodorous, glistening, scaly crystals, insoluble in water. Of all the members of the group, it most rarely produces toxic symptoms.

Treatment.—As for the other members of this group.

Exalgin, Aspirin, etc., as well as the above, may all act as poisons to certain persons, and even small medicinal doses may cause serious and even fatal consequences.

Symptoms (more or less common to all).—Nausea, vomiting, hurried respiration, marked cyanosis, syncope. Persistent sneezing and widespread urticaria may be present; collapse.

Aniline is an oily liquid, heavier than, and not soluble in, water. It is colourless or reddish-brown; it has a peculiar tar-like odour; it is soluble in alcohol, and forms a soluble sulphate with sulphuric acid. A solution of bleaching-powder gives with solution of the sulphate a purple colour changing to red-brown.

Symptoms.—Nausea, vomiting, giddiness, intoxication, drowsiness, gasping for breath, feeble pulse, and marked cyanosis. In its industrial use it may act as a poison either by inhalation of the fumes or by absorption through the skin. The symptoms then are mainly those of peripheral neuritis with blindness.

Fatal Dose.—About 6 drachms.

Treatment.—Wash out stomach; stimulants, artificial respiration, inhalation of oxygen, transfusion.

Nitro-benzol (Artificial Oil of Bitter Almonds).—It is used in perfumery, but is very poisonous when swallowed, or inhaled, or absorbed through skin. It is used in the manufacture of aniline dyes, and may act as an industrial poison. The symptoms closely resemble those of aniline poisoning, but there is perhaps greater mental confusion.

Fatal Dose.—Eight to ten drops have caused death.

Treatment.—Emetics, stimulants, transfusion of saline or blood, pituitrin, strychnine, or digitalin hypodermically.

Nitroglycerine gives rise to intense and persistent headache ('powder headache'). Throbbing and pulsation of all the arteries in the body; flushing of the face and collapse may follow.

Dinitrobenzene causes symptoms resembling nitro-benzol poisoning, and when acting as a chronic poison gives rise to weakness, jaundice, peripheral neuritis.

XXXVIII

SULPHONAL, TRIONAL, TETRONAL, VERONAL, PARALDEHYDE

These are dangerous drugs. The ordinary symptoms of the group are noises in the ears, headache, vertigo, inability to stand or to walk properly, insensibility, and cyanosis.

The most interesting point is the condition of the urine. In cases of poisoning it is dark or reddish-brown in colour, due to the presence of hæmatoporphyrin. It contains albumin and casts, but no red corpuscles. In cases of hæmatoporphyrinuria the prognosis is bad, and it is said that these cases invariably end fatally.

Treatment.—In an ordinary case emetics, strong coffee, hypodermic injections of strychnine, saline injections, and transfusion.

Cases of chronic poisoning from the 'als' are not uncommon, and are increasing in frequency. Hypnogen is apparently identical with veronal.

All the above-named aniline derivatives are included in Part I. of the scheduled poisons.

XXXIX

CONIUM AND CALABAR BEAN

Conium Maculatum (Spotted Hemlock).—All parts of the plant are poisonous, often mistaken for parsley. Contains the poisonous principle coniine, a volatile liquid alkaloid with a mousy smell; insoluble in water; soluble in alcohol, ether, and chloroform. It also contains methyl coniine.

Symptoms.—Dryness of throat, headache, dilated pupil, dysphagia, loss of muscular power, passing into complete paralysis. Delirium, coma, and convulsions, occasionally.

Post-Mortem Appearances.—Congested brain and lungs; redness of the mucous membrane of the stomach. The stomach and intestines should be examined for fragments of the leaves and fruit, recognized by their microscopical appearances.

Treatment.—Emetics, tannic acid or gallic acid. Diffusible stimulants.

Method of Extraction from the Stomach.—Use Stas-Otto process.

Tests.—The mousy odour. Deepened colour and dense white fumes with nitric acid. Pale red, deepening, with hydrochloric acid.

There are several other umbelliferous plants which are poisonous. The water hemlock (Cicuta virosa) produces symptoms not unlike those of hemlock; it has been mistaken for parsnip and celery. It contains an active principle, cicutoxin, which in some respects is allied to strychnine and picrotoxin. The fool's parsley, or lesser hemlock (Æthusa cynapium), is another member of this group, although doubt has been expressed as to whether it is really poisonous. The water dropwort (Œnanthe crocata) is undoubtedly poisonous, especially to cattle. In man it produces abdominal pain with diarrhœa and vomiting; dilated pupils, slow pulse, and cyanosis; delirium, insensibility, and convulsions. The post-mortem appearances are not characteristic, but the stomach and intestines should be examined for portions of the plant.

Calabar Bean or Physostigma.—The bean of Physostigma venenosum contains the alkaloid physostigmine or eserine, with the antagonistic alkaloid calabarine.

Symptoms.—Vomiting, giddiness, irregular cardiac action, contraction of the pupils, paralysis of lower extremities, and death from asphyxia.

Treatment.—Emetics; hypodermic injection of 1/50 grain sulphate of atropine, repeated if necessary.

Method of Extraction from the Stomach.—Use Stas-Otto process.

Test.—The contraction of the pupil which it causes.

XL

TOBACCO AND LOBELIA

Tobacco.—Nicotiana tabacum owes its poisonous properties to its alkaloid nicotine, a volatile, oily, amber-coloured liquid, with an acrid taste and ethereal odour; soluble in water, alcohol, ether, and chloroform. The drug has an intense depressant action on the heart and respiratory centre.

Symptoms.—Giddiness, fainting, nausea, and vomiting, with syncope, muscular tremors, stupor, stertorous breathing, and insensible pupil. Death has occurred after seventeen or eighteen pipes at a sitting.

Post-Mortem Appearances.—Not uniform or characteristic. General relaxed condition of muscles; engorgement of cerebral and pulmonary vessels. Congestion of gastric mucous membrane.

Treatment.—Emetics, stimulants, hypodermic injection of 1/25 grain of strychnine. Warmth to the surface by hot bottles, hot blankets.

Method of Extraction from the Stomach.—Digest the contents of the stomach in cold distilled water and very dilute sulphuric acid; strain, filter, and press residue. Evaporate the filtrate to half its bulk, digest with alcohol, and evaporate alcohol off in a water-bath. Dissolve residue (sulphate of nicotine) in water, and make solution alkaline with potash; then shake with ether in a test-tube. Remove ether and allow it slowly to evaporate. Test resulting alkaloid.

Tests.—No change of colour with the mineral acids. White deposit with corrosive sublimate. Sulphuric acid and bichromate of potassium give a green colour, oxide of chromium. Precipitate with bichloride of platinum and with carbazotic acid.

Lobelia Inflata (Indian Tobacco).—Much used in America by the Coffenite practitioners, and a valuable remedy for asthma.

Symptoms.—Nausea, vomiting, giddiness, cold sweats, prostration. Headache, giddiness, tremors, insensibility, and convulsions.

XLI

HYDROCYANIC ACID

Prussic Acid is the most active of poisons. The diluted hydrocyanic acid of the Pharmacopœia contains 2 per cent. of hydrocyanic acid, Scheele's 4 per cent. It is a colourless liquid, feebly acid, with odour of bitter almonds.

Cyanide of Potassium is largely used in photography and in electro-plating, and is also poisonous. It often contains undecomposed carbonate of potassium, which may act as a corrosive poison and cause erosion of the mucous membranes of the lips, mouth, and stomach.

Oil of Bitter Almonds, used as a flavouring agent, may contain (when improperly prepared) from 5 to 15 per cent. of the anhydrous acid.

Symptoms.—The symptoms usually come on in a few seconds, and are of the shortest possible duration. There is a sudden gasp for breath, possibly a loud cry, and the patient drops down dead. If the fatal termination is prolonged for a few minutes, the symptoms are intense giddiness, pallor of the skin, dilatation of the pupils, laboured and irregular breathing, small and frequent pulse, followed by insensibility. There may be convulsions or tetanic spasms, with evacuation of urine and fæces. Death results from paralysis of the central nervous system, but artificial respiration is useless, as the drug promptly arrests the heart's action. It also kills the protoplasm of the red blood-corpuscles, rendering them useless as oxygen-carriers.

Post-Mortem Appearances.—Skin livid, pale, or violet, with bright red patches on the dependent parts. The gastro-intestinal mucous membrane is bright red in colour, owing to the presence of cyanmethæmoglobin. Hands clenched, nails blue, jaws fixed, froth about mouth. Eyes prominent and glistening, odour of acid from body, venous system gorged.

Treatment.—Empty the stomach by the tube at once, and wash it out with a solution of sodium thiosulphate. Strong ammonia to the nostrils. Stimulants freely—brandy, chloric ether, ammonia, sal volatile ad libitum. If patient cannot swallow, inject hypodermically either brandy or ether. Hypodermic injection of 1/50 grain atropine. Douche to the face, alternately hot and cold. Death commonly occurs so rapidly that there is no time for treatment.

Fatal Dose (Smallest).—Half a drachm of the B.P. acid, equal to 0.6

grain of the anhydrous. Recovery from 1/2 ounce of the B.P. acid. These records are fallacious, for in specimens the percentage of anhydrous acid varies enormously. Practically, 1 grain of the anhydrous acid is fatal.

Fatal Period.—From two to five minutes after a large dose, but may be less.

Method of Extraction from the Stomach.—Having previously carefully fitted a watchglass to a wide-mouthed bottle, nearly fill the bottle with the contents of the stomach, blood, secretions, etc. Place a few drops of a solution of nitrate of silver on the concave surface of the watchglass, and cover the mouth of the bottle with it. The vapour of hydrocyanic acid, if present, will form a white precipitate which may be tested. Other watchglasses, treated with sulphide of ammonium or sulphate of iron and liquor potassæ, will give the reactions of the acid with appropriate tests. This method removes all objections as to foreign admixture. If the acid is not at first detected, gentle warming of the bottle in a water-bath will assist the evolution of the vapour. The vapour may be obtained by distillation, but this process is open to objections to which the other is not. In some cases it becomes changed in the body into formic acid, which should therefore be sought for.

Tests.—With nitrate of silver a white precipitate, insoluble in cold, but soluble in boiling, nitric acid. The precipitate heated, evolves cyanogen, having an odour of peach-blossoms, and burning, when lighted, with a pink flame. Liquor potassæ and sulphate of iron give a brownish-green precipitate, which turns to Prussian blue with hydrochloric acid. Liquor potassæ and sulphate of copper give a greenish-white precipitate, becoming white with hydrochloric acid. Sulphide of ammonium gives sulpho-cyanide of ammonium. This develops a blood-red colour with perchloride of iron, bleached by corrosive sublimate.

XLII

ACONITE

Aconite (Aconitum Napellus, monkshood).—Root and leaves. Poisonous property depends upon an alkaloid, aconitine. Aconite is one of the constituents of St. Jacob's Oil.

Symptoms.—Numbness and tingling in mouth, throat, and stomach, giddiness, loss of sensation, deafness, dimness of sight, paralysis, first of the lower and then of the upper extremities, vomiting, and shallow respiration. Pupils dilated. Pulse small, irregular, finally imperceptible. The mind remains unaffected. Death often sudden.

Post-Mortem Appearances.—Venous congestion, engorgement of brain and membranes.

Treatment.—Emetics, stimulants freely. Best antidote is sulphate of atropine, 1/50 grain hypodermically, and also strychnine. Digitalis also useful. Warmth to whole body. Patient to make no exertion.

Fatal Dose.—Of root or tincture, 1 drachm.

Fatal Period.—Average, less than four hours.

Method of Extraction from the Stomach, etc.—Extraction from contents of stomach by Stas-Otto process. It may be found in the urine; gives usual alkaloidal reactions, but no distinctive chemical test known.

Tests.—Chiefly physiological; tingling and numbness when applied to tongue or inner surface of cheek. Effects on mice, etc. A cadaveric alkaloid or ptomaine has been found in the body, possessing many of the actions of aconitine. The presence of this substance was suggested in the Lamson trial.

The Indian aconite, Aconitum ferox, the Bish poison, is much more active than the European variety. It contains a large proportion of pseudaconitine, and is frequently employed in India, not only for the destruction of wild beasts, but for criminal purposes.

Aconitine varies much in activity according to its mode of preparation and the source from which it is derived. The most active kind is probably made from A. ferox.

XLIII

DIGITALIS

All parts of the plant Digitalis purpurea (purple foxglove) are poisonous. Contains the glucoside digitalin and other active principles.

Symptoms.—Nausea, vomiting, purging, and abdominal pains. Vomited matter grass-green in colour. Headache, giddiness, and loss of sight; pupils dilated, insensitive; pulse weak, remarkably slow and irregular; cold sweat. Salivation occasionally, or syncope and stupor. Death sometimes quite suddenly.

Post-Mortem Appearances.—Congested condition of brain and membranes; inflammation of gastric mucous membrane.

Treatment.—Emetics freely; infusions containing tannin, as coffee, tea, oak-bark, galls, etc. Stimulants. Hypodermic injection of 1/120 grain of aconitine.

Method of Extraction from the Stomach, etc.—Use Stas-Otto process.

Tests for Digitalin.—A white substance, sparingly soluble in water, not changed by nitric acid; turns yellow, changing to green, with hydrochloric acid. The minutest trace of digitalin moistened with sulphuric and treated with bromine vapour gives a rose colour, turning to mauve. This is very delicate, but in experienced hands the physiological test is more reliable. The chemist who has had no practical experience in pharmacological methods would be wiser to keep to his chemical tests.

XLIV

NUX VOMICA, STRYCHNINE, AND BRUCINE

Nux Vomica consists of the seeds of the Strychnos nux vomica. From these strychnine and brucine are obtained. The symptoms, post-mortem appearances, and treatment, of poisoning by nux vomica are the same as for strychnine.

Strychnine is a powerful poison, and forms the active ingredient of many 'vermin-killers.' It occurs as a white powder or as colourless crystals, with a persistent bitter taste; very slightly soluble in water; more or less soluble in benzol, ether, and alcohol.

Symptoms.—Sense of suffocation, twitchings of muscles, followed by tetanic convulsions and opisthotonos, each lasting half to two minutes. Mental faculties unaffected, face congested and anxious; eyes staring, lips livid; much thirst. The period of accession of the symptoms varies with the mode of administration of the poison. Symptoms, as a rule, come on soon after food has been taken. Patient may die within a few hours from asphyxia or from exhaustion.

In Tetanus there is usually history of a wound; the symptoms come on slowly; lockjaw is an early symptom, and only later complete convulsions; the intervals between the fits are never entirely free from rigidity. Death is delayed for some days.

Post-Mortem Appearances.—Heart empty, blood fluid, rigor mortis persistent. Hands usually clenched; feet arched and inverted. Congestion of brain, spinal cord, and lungs.

Treatment.—Emetics or stomach-pump if the patient is deeply anæsthetized. Tannic acid and permanganate of potassium. Bromide of potassium 1/2 ounce with chloral 30 grains, repeated if necessary.

Fatal Dose (Smallest).—Quarter of a grain.

Fatal Period (Shortest).—Ten minutes; usually two to four hours.

Method of Extraction from the Stomach.—The alkaloid may be separated by the process of Stas-Otto.

Tests.—Strychnine has a characteristic, very bitter taste; it imparts this taste to even very dilute solutions; it is unaffected by sulphuric acid, but gives a purple-blue colour, changing to crimson and light red, when the edge of this solution is touched with dioxide of manganese, potassium bichromate, ferricyanide of potassium, or permanganate of potassium. This test is so delicate as to show the 1/25000 of a grain of the alkaloid. A very minute quantity (1/5000 grain) in solution placed on the skin of a frog after drying causes tetanic convulsions.

Brucine.—This alkaloid, found associated with strychnine, possesses the same properties, though in a less powerful degree. Nitric acid gives a blood-red colour, changed to purple with protochloride of tin.

XLV

CANTHARIDES

Cantharides.—Spanish fly, or blistering beetle, is the basis of most of the blistering preparations. It is sometimes taken as an abortifacient or given as an aphrodisiac, but whether it has any such action is open to question. It acts as an irritant to the kidneys and bladder, and sometimes produces haæmaturia and a good deal of temporary discomfort.

Symptoms.—Burning sensation in the throat and stomach, with salivation, pain and difficulty in swallowing. Vomiting of mucus mixed with blood. Tenesmus, diarrhœa, the motions containing blood and mucus. Dysuria, with passage of small amounts of albuminous and bloody urine. Peritonitis, high temperature, quick pulse, headache, loss of sensibility, and convulsions.

Post-Mortem.—Gastro-intestinal mucous membrane inflamed, with gangrenous patches. Genito-urinary tract inflamed. Acute nephritis.

Treatment.—An emetic of apomorphine; demulcent drinks, such as barley-water, white of egg and water, linseed-tea and gruel (but not oils), with a hypodermic injection of morphine to allay pain.

Tests.—The vomited matter often contains shining particles of the powder. The urine will probably be albuminous.

XLVI

ABORTIFACIENTS

Emmenagogues are remedies which have the property of exciting the catamenial flow; ecbolics, or abortives, are drugs which excite contraction of the uterus, and are supposed to have the power of expelling its contents. The vegetable substances commonly reputed to be abortives are ergot, savin, aloes (Hierapicra), digitalis, colocynth, pennyroyal, and nutmeg; but there is no evidence to show that any drug possesses this property. Lead in some parts of the country is a popular abortifacient. A medicine may be an emmenagogue without being an ecbolic. Permanganate of potassium and binoxide of manganese are valuable remedies for amenorrhœa, but will not produce abortion. The vegetable substances frequently used as abortives are savin and ergot.

Savin (Juniperus Sabina).—Leaves and tops of the plant yield an acrid oil having poisonous properties, and which has even produced death.

Symptoms.—Those of irritant poisons. Purging not always present, but tenesmus and strangury.

Post-Mortem Appearances.—Acute inflammation of alimentary canal. Green powder found. This, washed and dried and then rubbed, gives odour of savin.

Test.—A watery solution of savin strikes deep green with perchloride of iron, and if an infusion of the twigs has been taken the twigs may be detected with the microscope. The twigs obtained from the stomach, dried and rubbed between the finger and thumb, will give the odour of savin.

Ergot (Secale Cornutum).—A parasitic fungus attacking wheat, barley, oats, and rye, which is reputed to have the power of causing contraction of unstriped muscular fibre, especially that of the uterus.

Symptoms.—Lassitude, headache, nausea, diarrhœa, anuria, convulsions, coma. Small quantities frequently repeated have in the past produced gangrene of the extremities, or anæsthesia of fingers and toes.

Tests.—Lake-red colour with liquor potassæ; this liquid filtered gives a precipitate of same colour with nitric acid.

XLVII

POISONOUS FUNGI AND TOXIC FOODS

Fungi.—Of the poisonous mushrooms, the Amanita phalloides and the fly agaric, or Agaricus muscarius, are the most potent. The active principle of the former is phallin, and of the latter muscarine. The Amanita phalloides is distinguished from the common mushroom (Agaricus campestris) by having permanent white gills and a hollow stem. The Agaricus muscarius is bright red with yellow spots. Phallin is a toxalbumin which destroys the red blood-corpuscles, causing the serum to become red in colour and the urine blood-stained. Fibrin is liberated, and thromboses occur, especially in the liver. The symptoms may be mistaken for phosphorus-poisoning or acute yellow atrophy of the liver. Muscarine affects the nervous system chiefly.

Edible fungi have an agreeable taste and smell, and are firm in substance. Poisonous fungi have an offensive smell and bitter taste, are often of a bright colour, and soon become pulpy.

Symptoms.—These may be of the narcotic or irritant types. Usually, however, there is violent colic, with thirst, vomiting, and diarrhœa, mental excitement, followed by delirium, convulsions, coma, slow pulse, stertorous breathing, cyanosis, cold extremities, and dilated pupils.

Post-Mortem.—In phallin-poisoning the blood remains fluid; numerous hæmorrhages are present, with fatty degeneration of the internal organs.

Treatment.—Use the stomach-tube to give a solution of permanganate of potash, emetics, followed by a hypodermic injection of 1/50 grain of atropine. Transfusion of saline fluid. A dose of castor-oil would be useful.

Foods.—The kinds of food which most frequently produce symptoms of poisoning are pork, veal, beef, meat-pies, potted and tinned meats, sausages, and brawn. Sausage-poisoning is common in Germany. It is not necessary that the food should be 'high' to give rise to poisoning. It may arise from the use of the flesh of an animal suffering from some disease, from inoculation with micro-organisms, or from the presence of toxalbumoses or ptomaines. Many diseases, such as diarrhœa, enteric fever, and cholera, and perhaps tuberculosis, may be caused by eating infected food. Trichiniasis may also be mentioned. Tinned fish often gives rise to symptoms of poisoning, and shell-fish are not uncommonly contaminated with pathogenic micro-organisms. Mussel-poisoning was formerly supposed to be due to the copper in them derived from ships' bottoms, but it is more probably the result of the formation of a toxine during life, and not after decomposition has set in. Milk, too, may give rise to gastro-intestinal irritation from the occurrence in it of chemical changes. There have been epidemics of poisoning from eating cheese containing tyrotoxicon. Ergotism from eating bread made with ergotized wheat is now rare, but pellagra from the consumption of mouldy maize, and lathyrism, due to the admixture with flour of the seeds of certain kinds of vetch, are still common in Southern Europe.

Symptoms.—The symptoms which result from the ingestion of poisonous meat are often very severe. In some cases their appearance is delayed from twenty-four to forty-eight hours. They may resemble those of an infectious disease or those of acute enteritis. Usually there are headache, anorexia, rigors, intestinal disturbance, pains in the back and limbs, and delirium. Sometimes the symptoms resemble atropine-poisoning, a condition due to ptomatropine.

Treatment.—Emetics, purgatives, stimulants, with hypodermic injections of strychnine and atropine along with stimulants.

XLVIII

PTOMAINES OR CADAVERIC ALKALOIDS

Every medical man, before presenting himself to give evidence in a case of suspected poisoning, should make himself thoroughly acquainted with recent researches on the subject. Ptomaines are, for the most part, alkaloids generated during the process of putrefaction, and they closely resemble many of the vegetable alkaloids—veratrine, morphine, and codeine, for example—not only in chemical characters, but in physiological properties. They are probably allied to neurine, an alkaloid obtained from the brain and also from the bile. Some of them are analogous in action to muscarine, the active principle of the fly fungus. Some are proteids, albumins, and globulins. Ptomaines may be produced abundantly in animal substances which, after exposure under insanitary conditions, have been excluded from the air. Ptomaines or toxalbumins are sometimes found in potted meats and sausages, and are due to organisms—the Bacillus botulinus, the B. enteritidis of Gärtner, the B. proteus vulgaris, or the B. ærtrycke (which is perhaps the most common of all). The symptoms produced by the latter are usually vomiting, abdominal pain, pains in the limbs and cramps, diarrhœa, vertigo, coldness, faintness, and collapse. The symptoms of botulism are dryness of skin and mucous membranes, dilatation of pupils, paralysis of muscles, diplopia, etc. Articles of food most often associated with poisoning are pork, ham, bacon, veal, baked meat-pie, milk, cheese, mussels, tinned meats.

In a case of suspected poisoning, counsel for the defence, if he knows his work, will probably cross-examine the medical expert on this subject, and endeavour to elicit an admission that the reactions which have been attributed to a poison may possibly be accounted for on the theory of the formation of a ptomaine. There is practically no counter-move to this form of attack.

www.ingramcontent.com/pod-product-compliance
Lightning Source LLC
Chambersburg PA
CBHW011255040426
42453CB00015B/2410